Computerized Tomography for Scientists and Engineers

Computerized Tomography for Scientists and Engineers

Edited by
PRABHAT MUNSHI
Department of Mechanical Engineering
Indian Institute of Technology Kanpur
Kanpur-208 016, India

CRC is an imprint of the Taylor & Francis Group,
an informa business

Anamaya Publishers
New Delhi

EDITOR
Prabhat Munshi
Professor
Department of Mechanical Engineering
Indian Institute of Technology
Kanpur-208 016, India

Copublished by CRC Press
Taylor and Francis Group
6000 Broken Sound Pkwy, NW
Suite 300, Boca Raton FL 33487, USA
with Anamaya Publishers, New Delhi, India

ISBN 1-4200-4793-0

Sold and distributed exclusively in all the countries, except India, by CRC Press
6000 Broken Sound Pkwy, NW, Suite 300, Boca Raton FL 33487, USA

In India, sold and distributed by Anamaya Publishers
F-154/2, Lado Sarai, New Delhi-110 030, India

All rights reserved. No part of this publication may be reproduced, stored in a retrieval system or transmitted in any form or by any means, electronic, mechanical, photocopying or otherwise, without the prior permission of the copyright owner.

This book contains information obtained from authentic and highly regarded sources. Reprinted material is quoted with permission, and sources are indicated. Reasonable efforts have been made to publish reliable data and information, but the editors and the publishers cannot assume responsibility for the validity of all materials or for the consequences of their use.

Copyright © 2007 Anamaya Publishers, New Delhi, India

Printed in India.

Foreword

Computerized tomography (CT) is a non-invasive technique for imaging 3D objects. It has been introduced in medical radiology in the early 1970s. Since then, many competing techniques, such as magnetic resonance imaging (MRI), nuclear medicine (PET and SPECT) and ultrasound transmission tomography came up. One of the most exciting developments of the last centuries is the transition of these medical imaging techniques to other fields of science, such as non-destructive testing, process or industrial tomography, radar and seismic imaging.

The present volume gives examples of applications of tomography to engineering. Typical problems are monitoring of multiphase flows, crystal growth, blast furnaces and stirred vessels, non-destructive testing, plasma diagnostics, and determining the strength of bones. The techniques used comprise X- and γ-rays, electrical impedance and resistance measurements, ultrasound, and lasers. Various mathematical issues are addressed, for example: reconstruction algorithms in general, problems with few data (often only four projections can be taken), vector tomography, inverse scattering in the Born approximation. Various physical problems (such as beam hardening) are discussed.

The book provides an excellent account of the present lively developments in imaging in engineering. Also, the reader will notice that the same problems come up in quite diverse fields, and that identical mathematical techniques can be used. This establishes imaging as a truly interdisciplinary field, with mathematics as the common language.

Prof. Dr. Dr. h.c. Frank Natterer
Department of Mathematics
University of Münster
Germany

Preface

Applications of computerized tomography (CT) have been growing in the non-medical areas for the past 20 years. Indian Institute of Technology, Kanpur has provided a launching pad for several researchers across the country to try new ideas and applications. Initial work was done in error analysis of algorithms and numerical simulation. It has slowly expanded now to significant experimental activities in the Department of Mechanical Engineering, IIT Kanpur. As a result, courses on tomography as well as non-destructive testing are regularly being offered by the department to students of nuclear, mechanical, chemical, civil, aerospace and electrical engineering.

Computerized tomography is a non-invasive technique for imaging 3D objects. It has been introduced in medical radiology in the early 1970s. Since then, many competing techniques, such as magnetic resonance imaging (MRI), nuclear medicine (PET and SPECT) and ultrasound transmission tomography came up. One of the most exciting developments of the last centuries is the transition of these medical imaging techniques to other fields of science, such as non-destructive testing, process or industrial tomography, radar and seismic imaging.

The present volume gives examples of applications of tomography to engineering. Typical problems are monitoring of multiphase flows, crystal growth, blast furnaces and stirred vessels, non-destructive testing, plasma diagnostics, and determining the strength of bones. The techniques used comprise X- and γ-rays, electrical impedance and resistance measurements, ultrasound, and lasers. Various mathematical issues are addressed, for example: reconstruction algorithms in general, problems with few data (often only four projections can be taken), vector tomography, inverse scattering in the Born approximation. Various physical problems (such as beam hardening) are discussed.

The volume provides an excellent account of the present lively developments in imaging in engineering. Also, the reader will notice that the same problems come up in quite diverse fields, and that identical mathematical techniques can be used. This establishes imaging as a truly interdisciplinary field, with mathematics as the common language.

The editor wishes to acknowledge useful discussions held with Dr. Baldev Raj (IGCAR Kalpakkam), Dr. P. Satyamurthy and Dr. V.K. Wadhawan (BARC Mumbai), Dr. C. Muralidhar (DRDL Hyderabad), Dr. S. Vathsal, ER&IPR New Delhi and Dr. K. Balasubramaniam (IIT Madras).

The editor gratefully acknowledges his collaboration with Dr. N.N. Kishore and Dr. K. Muralidhar, his colleagues at IIT Kanpur, who have spearheaded the CT activities to a very exciting and meaningful conclusion. Continuous encouragement and support of Prof. S.G. Dhande, Director, IIT Kanpur, is sincerely appreciated.

The editor is indebted to Prof. Frank Natterer who inspired the CT activities almost 20 years ago at IIT Kanpur and is still a constant source of help and inspiration. His participation in the first CT Tutorial (CT2001) was a great morale booster for the Indian CT community who learned very quickly that it is equally important to give CT images with precise error bands.

Financial support received from Department of Mechanical Engineering, IIT Kanpur, for publication of this volume is gratefully acknowledged. The editor thanks all the authors, who are experts in their fields, for their contribution in this volume. A special note of thanks to the members of Tomography Laboratory (R.K. Jauhari, Dr. A.M. Quraishi, Mayuri Razdan) for their untiring efforts.

Comments and suggestions from the readers are welcome.

EDITOR

Contents

Foreword   v
Preface   vii

1. Process Tomography: Development and Application of Non-intrusive Measuring Techniques for Multiphase Flows   1
   M. Behling and D. Mewes

2. Impedance Technique for the Measurement of Two Phase Flow Parameters: Possibilities and Challenges   10
   P.K. Das, G. Das, S. Sen and K. Biswas

3. X-Ray Computed Tomography for Solid Objects   27
   A.K. Jena, N.K. Das, S.J. George, B. Venkataraman, C. Babu Rao, K. Kasiviswanathan, T. Jayakumar, P. Kalayanasundaram and Baldev Raj

4. Steady-State Multi-Phase Flow Measurement Facility at FCRI   34
   M. Suresh, R.V. Rajesh, M. Viswanathan and M.S. Konnur

5. Bone Imaging Using Compound Ultrasonic Tomography   48
   P. Lasaygues and P. Laugier

6. Challenges in Quantitative Ultrasound Bone Strength Assessment: Status and Perspectives   57
   P. Laugier

7. Computerized Tomography in Blast Furnace   68
   S.K. Mandal

8. Electrical Process Tomography: Imaging Fluid Mixing Processes Inside Stirred Vessels   75
   R. Mann

9. Convection in Differentially Heated Fluid Layers and Its Reconstruction Using Radial Tomography in an Octagonal Cavity   87
   Sunil Punjabi

10. Tomography in Fusion Plasma Research   106
    C.V.S. Rao

11. Tomographic Reconstructive Techniques for Void Fraction Distribution in Heavy Density Liquid Metal Two-phase Flows   118
    P. Satyamurthy, N.S. Dixit, R. Chaudhary and P. Munshi

12. Imaging of Buoyancy-Driven Convective Field Around a KDP Crystal Using
    Schlieren Tomography   133
    Atul Srivastava, K. Muralidhar and P.K. Panigrahi

13. Development of Computer Aided Tomography Systems in DRDL   148
    S. Vathsal, C. Muralidhar, G.V. Siva Rao, K. Kumaran, M.P. Subramanian,
    M.R. Vijaya Lakshmi, Sijo N. Lukose and M. Venkata Reddy

14. Determination of the Concentration Field Around a Growing Crystal Using
    Laser Shadowgraphic Tomography   158
    Sunil Verma, K. Muralidhar and V.K. Wadhawan

15. Digital Radiography for Non-destructive Testing   176
    Debasish Mishra, Rajashekar Venkatachalam and V. Manoharan

16. 3D Tomography Using Neutrons and X-Rays   214
    Amar Sinha and P.S. Sarkar

    INDEX   227

Computerized Tomography for Scientists and Engineers

Computerized Tomography for Scientists and Engineers
Edited by P. Munshi
Anamaya Publishers, New Delhi, India

1. Process Tomography: Development and Application of Non-intrusive Measuring Techniques for Multiphase Flows

M. Behling and D. Mewes

University of Hannover, Institute of Process Engineering, IfV
Callinstrasse 36, 30167 Hannover, Germany

Abstract

The variety of ready to use measuring techniques available in the market is ever growing. But nevertheless in multiphase flows the non-intrusive acquisition of local measurements of process parameters like temperature, concentration or volume fractions often is still a challenge. In many cases, instead of the desired local measurements only integral measurements, for instance, along lines, can be obtained. This is the point where process tomography comes into play. In general, in tomography the desired local values of the respective properties are reconstructed mathematically from a certain number of integral measurements. Depending on the measurement task, a variety of optical, electrical conductance, electrical capacitance, X-ray, and some other techniques can be used. The various techniques differ significantly in their suitability for different measurement objects, spatial and temporal resolution, availability in the market, cost, and many more parameters. In process engineering research the desired tomographic systems often are still not offered commercially. The Institute of Process Engineering, University of Hannover, Germany, looks back on many years of developing, building and using its own tomographic systems, covering all the abovementioned techniques. This article gives an overview of the tomographic measuring systems built and used at the Institute of Process Engineering and the research work they applied for.

1. Introduction

Tomographic measuring techniques have their origin back in the 1960s in medical applications. The first tomographic measuring systems ever built were X-ray computer tomographs (CT) for scanning the human skull, later for scanning the whole human body. Still today, medical X-ray CT is the most widespread application of tomography, followed by medical nuclear magnetic resonance (NMR) tomography. But meanwhile, both X-ray and NMR tomography as well as a broad variety of other tomographic techniques found their way into various technical fields of application, both in non-destructive material testing as well as in non-invasive flow measurement. Tomographic techniques can be based on many different physical measurements, like attenuation of X-rays, attenuation of light, electrical conductivity, electrical capacity, and many more. All tomographic techniques have several advantages in common. In general, they are non-destructive and non-invasive. While, for instance, probes held into fluid flows influence and change the flow to be measured, tomographic techniques do not. Another advantage is that while a probe gives a local measurement at only one point at a time, tomographic measurement results are whole cross-sections at one time. Tomographic techniques provide local measurements while without tomography in many cases only integral measurements across the measured object or apparatus can be obtained. Thus, tomographic techniques have the potential to reveal details no other measurement technique can. In medical tomography, systems are produced in high numbers and sold as turnkey systems. For some technical applications this is also the case, like for instance X-ray tomography of metal castings. But in process tomography the required measurement techniques normally are not available in the market as turnkey systems. Instead the user often is forced to build and put together his own systems. This is the reason why the Institute of Process Engineering, Hannover, Germany, puts lots of effort into developing these techniques.

2. Tomographic Measurement Principle

The principle of tomographic measurement technique shown in Fig. 1 is same for all tomographic techiques, irrespective of whether it is medical or technical application or based on X-ray attenuation or electrical capacity. Tomography always is a two-step process. The first step is the physical measurement itself. With the object in the sensor, with some sort of measuring device, a certain number of integral measurements is taken. These measurements can be based on various physical effects. Some of these are presented in more detail later in this article.

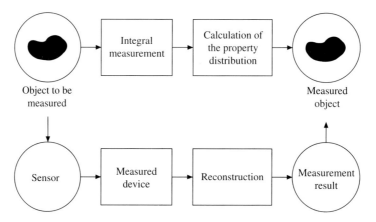

Fig. 1 Tomographic measurement principle.

After collecting a certain number of these integral measurements, the second step in tomography is the so-called reconstruction. In this step, the measured raw data is used to calculate the distribution of the measured property over the whole cross-section through the measured object, giving a two-dimensional slice with local measurement results. The various reconstruction techniques that can be applied are not within the scope of this article.

The sets of integral measurements, for instance line integrals along parallel X-ray or laser beams, are often called *projections*. Fig. 2 shows schematically an example with a set of three projections, each consisting of parallel line integrals. The single values of Φ_M are linearly independent integrals of the local physical property $g(x, y)$ in the measurement plane. The complete measurement vector $\vec{\Phi}_M$ (consisting of all the projections) intrinsically contains the full information about $g(x, y)$, but not explicitly. Therefore, the reconstruction of $g(x, y)$ can be a rather delicate process.

The reconstructed property $g(x, y)$ depends on the physical measurement principle the tomographic technique is based on. X-ray tomography gives the distribution of the X-ray attenuation coefficient, electrical capacitance tomography gives the distribution of the electrical permittivity over the cross-section. These properties can in turn be translated into phase distributions in

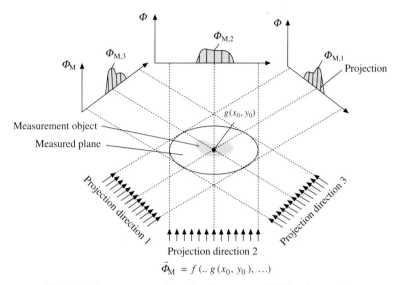

Fig. 2 Measurement of projections—parallel line integrals.

multiphase flow, when the X-ray attenuation coefficient or the electrical permittivity of water and air differ from each other. In process tomography, very often the phase distribution in the measured cross-section is the desired result.

Tomographic techniques can be based on a variety of physical effects which are exploited for obtaining the integral measurements mentioned above. This results in differences in temporal and spatial resolution. Fig. 3 provides only a schematic overview, since not all different existing systems can be taken into account in detail. For instance, in spatial resolution significant differences exist between standard X-ray and microfocus X-ray techniques. In general, techniques that do not require any moving parts (like electrical tomography) can collect measurements much faster than those techniques based on, e.g. a mechanical rotation (like X-ray tomography). For any application, the appropriate tomographic technique has to be chosen by taking into account

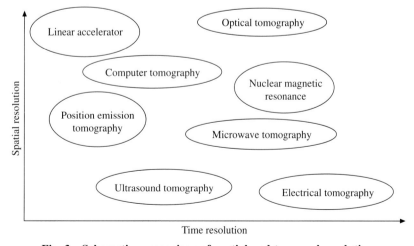

Fig. 3 Schematic comparison of spatial and temporal resolution.

physical constraints (like optical techniques cannot be applied to opaque objects) and the required resolution (i.e. the fluid flow quasi-stationary or highly transient?). In the experimental research work at the Institute of Process Engineering, Hannover, electrical, optical and X-ray tomography are used which are not fully available in the market in the desired configurations. Therefore, the Institute is active in developing and building its own systems. Within electrical tomography, both electrical conductivity as well as electrical capacitance tomography systems are used and built in Hannover. Following section gives a short introduction into X-ray tomography and the two types of electrical tomography.

3. X-ray Tomography

X-rays are electromagnetic radiation with a short wavelength and high photon energy and can penetrate any material to a certain extent. Along their path through the material, the X-rays are attenuated. The longer the path length in the material, higher is the material's density, and more radiation is attenuated. Therefore, measuring the intensity of the attenuated radiation allows to 'guess' what was within the path of the X-ray. In standard X-ray imaging (radiography), pure projection images are the final measurement results, while in X-ray tomography multiple projections are used to calculate cross-sections (also called slices) through the object.

The principle set-up of X-ray computer tomography (CT) is shown in Fig. 4. X-rays are generated by so-called X-ray tubes, which are point sources. Out of the generated radiation, only a fan beam is used. The measured object is irradiated by this fan beam. The intensity of the attenuated radiation is measured along a line with an X-ray detector, for instance, a special X-ray sensitive diode array. The object is rotated to collect many of these line integrals through the object in many different angles. In medical or industrial X-ray tomography upto several thousands of projections are taken. The set of projections, as explained above, is the basis for the reconstruction.

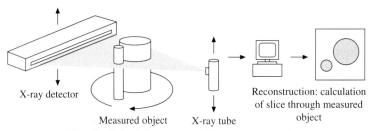

Fig. 4 **Principle set-up of X-ray tomography.**

For scanning multiple slices, the X-ray tube and detector are moved vertically. A complete stack of slices, put on top of each other, is a complete 3D representation of the measured object. The set-up described here is used in industrial systems for inspection of metal castings. Apart from the mechanism described here, various other mechanical set-ups are possible for collecting the projections. For instance, instead of rotating the object, the X-ray source and detector can be rotated around the measured object (used in medical tomography). Instead of using only a fan beam, a cone beam together with an area detector can be used. These various options are often referred to as 'generations'. The principle described above is the so-called '3rd generation'.

The Institute operates an industrial, commercially available X-ray tomography system. A photo of the system is shown in Fig. 5. The system features a 420 kV X-ray tube, fan beam collimator and a digital line detector capable of distinguishing 2^{18} intensity levels of the measured radiation. It was designed mainly for material testing and is perfectly suited for determining the exact geometry of parts and finding cracks, voids and inclusions in metal castings or in fibre composite materials. Such systems are mainly used in the aircraft and automotive industry. At the Institute of Process Engineering, Hannover, this X-ray CT is used for measuring cross-sections through the two-phase flow of water and air in columns with packed beds and structured packings. In these objects, under certain parameters of operation the fluid flow can be assumed to be quasi-stationary. Therefore, the relatively long measurement time of several seconds per slice can be accepted.

Additionally, during the rotation of the measured object the fluid flow in the structures will follow the rotation, provided the speed of rotation is slow enough. The advantage is the very high resolution of the measured images. Measurements of a packed bed and a structured packing are shown in Fig. 6. The higher the grey value in the image, the higher is the density of the material in the measured object. Black is air.

To extract the distribution of the fluid flow even more clearly, 'dry' images are subtracted from the 'wet' images. The remaining difference-image shows only the fluid itself. In Fig. 7, several of these difference-images taken at various superficial velocities of the liquid have been color-coded and overlaid.

Fig. 5   Industrial X-ray tomograph.

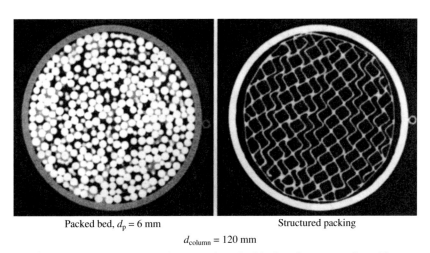

Packed bed, d_p = 6 mm   Structured packing

d_{column} = 120 mm

Fig. 6   X-ray tomography images of packed bed and structured packing.

The slices measured through such a column can be added to a full 3D view of the flow field. Fig. 8 shows the liquid film in a random packed bed. Only the liquid itself is shown, with the solid particles subtracted.

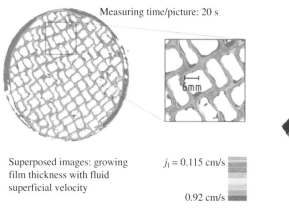

Fig. 7 Liquid distribution in structured packing.

Fig. 8 Liquid film in random packing with solid particles subtracted.

The set-up with the measured object rotating between X-ray source and sensor only is suited for objects that fully follow the rotation. It cannot be used for objects like a bubble column, where only the apparatus could be rotated, but not the fluid flow. For conducting measurements in a bubble column, at the Institute of Process Engineering, Hannover, a second X-ray CT system is set-up. Instead of rotating the object, the X-ray source and detector are rotated around the bubble column. The main limitation of X-ray CT is the relatively long measurement time per slice. Therefore, in measured objects that change very fast, only time averages can be measured. For every pixel in a reconstructed slice, the time averaged volume fractions of various phases are reconstructed.

4. Electrical Capacitance Tomography

Another technique that can be used for measurements in packed beds and structured packings (and also for other objects of course) is electrical capacitance tomography. The principle set-up of the sensor is shown in Fig. 9.

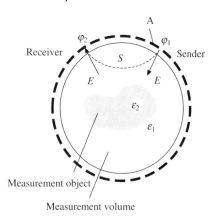

Fig. 9 Schematic set-up of electrical capacitance tomography sensor.

Around the circumference of the circular measurement volume (for instance a pipe or a column) a certain number of electrodes are mounted peripherically. Between any two of these electrodes, called sender and receiver, the electrical capacitance is measured. This capacitance depends on the relative position of the electrodes towards each other and on the material distribution, i.e. on the distribution of the permittivities ε. Each single measured capacity depends on all permittivities in the whole measurement volume, they all influence the measured integral value. Hence, such tomography is non-linear, in contrary to linear tomography techniques like X-ray CT. This non-linearity makes the reconstruction process much more complicated compared to the various linear tomography techniques. Thus, in electrical capacitance tomography very often iterative techniques are applied.

Compared to X-ray tomography, the spatial resolution of electrical capacitance tomography is much lower. It is not possible to improve the resolution by using a higher number of smaller electrodes too much, because the size of each electrode must not be too small in order to measure the resulting capacitances. The great advantage of capacitance tomography is its time resolution. It is possible to get several images within fraction of a second. Some example measurement results are shown in Fig. 10.

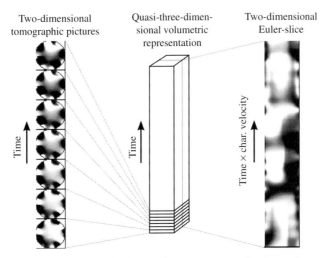

Fig. 10 Measurement results of electrical capacitance tomography (two-phase flow in trickle bed).

The measurements were taken in one slice in the two-phase flow in trickle bed through a random packing. The packing consists of ceramic spheres. On the left side of Fig. 10 the sequentially measured pictures are depicted. Regions with a high liquid hold-up are shown in dark color. The ceramic spheres are not visible, because they have been removed out of the measurement values by the calibration of the measurement system. The slices have a temporal distance of 3 ms and show the change of the phase distribution inside the sensor. They can be added to each other, that results in a quasi-three-dimensional representation of the measurement results. Through this a vertical slice extracts the temporal change of the phase distribution in the measurement plane.

5. Electrical Conductivity Tomography

This second electrical tomography technique is based on measuring electrical conductivities. The sensors used for this technique in Hannover consist of three sets of parallel wires, as shown in Fig. 11. Due to its construction, this type of sensor is also called 'wire-mesh' sensor.

Between each two parallel wires, the electrical conductivity is measured. This conductivity depends on the material between the wires. This technique is very useful for measuring the two-phase flow of water and air, for instance. In this case, the measured values of the conductivity are roughly proportional to the amount of water between the wires. When the conductivities for all pairs of neighboring wires in all the three measurement planes have been collected, the distribution in the flow field can be reconstructed. This technique is used for measuring two-

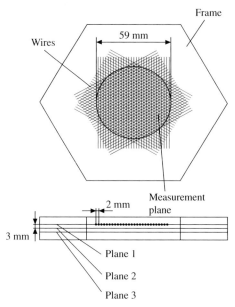

Fig. 11 Electrical conductivity tomography sensor.

phase flows of water and air in horizontal pipes, for instance. As an example, a sequence of slices through a water plug in two-phase flow is shown in Fig. 12.

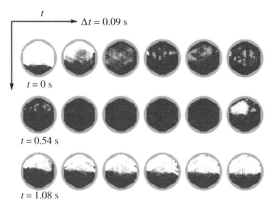

Fig. 12 Water plug in two-phase flow in horizontal pipe measured by electrical conductivity tomography.

This again is a linear tomography technique. The measured conductivity between two wires depends on the flow field between these wires and is (nearly) independent from the rest of the measurement plane. Like in electrical capacitance tomography, one advantage of electrical conductivity tomography is the very high time resolution, which makes this technique suitable for very fast fluid flow measurements. The spatial resolution is higher than that of capacitance tomography, and the required reconstruction algorithms are easier to handle. A disadvantage is that of course the wire mesh sensor is not completely non-invasive. While for highly turbulent fluid flows the influence of the wires can be neglected, this technique cannot be used when solids come into play.

6. Conclusions and Further Reading

X-ray tomography offers very high spatial resolution, while its time resolution is rather poor. Pixel sizes far below 1 mm are no problem, but measurement time ranges from a few seconds upto several minutes. A great advantage is that X-ray tomography is not restricted to certain materials or types of measured object. The electrical tomography techniques on the other hand offer a very high time resolution. Hundreds of images can be taken within a second. But the spatial resolution compared to X-ray tomography stays far behind. And the electrical techniques are much more limited to certain objects like multiphase flows in pipes and columns.

This article only gives a short and rough introduction. For further reading, several publications by Mewes et al. [1-6] are recommended. These publications offer more details about the tomographic techniques used and built in Hannover and the reconstruction algorithms which were not explained in this article. They also contain numerous references of publications by other groups also working in various fields of tomographic measurement techniques. For detailed information about the mathematical background of various reconstruction algorithms, especially the publications by Natterer [7], Herman [8] and Kak and Slaney [9] are recommended.

References

1. N. Reinecke, G. Petritsch, D. Schmitz and D. Mewes, Tomographische Messverfahren-Visualisierung zweiphasiger Strömungsfelder, *Chemie Ingenieur Technik*, **69** (1997) 10, 1379–1394.
2. C.G. Xie, N. Reinecke, M.S. Beck, D. Mewes and R.A. Williams, Electrical tomographic techniques for process engineering applications, *Chem. Eng. J.*, **56** (1995), 127–133.
3. D. Schmitz, N. Reinecke, G. Petritsch and D. Mewes, High resolution x-ray tomography for stationary multiphase flows; OECD/CSNI Specialist Meeting on Advanced Instrumentation and Measurement Techniques, Santa Barbara, CA, March 17–20, 1997.
4. N. Reinecke and D. Mewes, Recent developments and industrial/research application of capacitance tomography, *Measurement Science and Technology*, **7** (1996) 3, 233–246.
5. N. Reinecke and D. Mewes, Tomographic imaging of trickle-bed reactors, *Chem. Eng. Sci.*, **51** (1996) 10, 2131–2138.
6. N. Reinecke, M. Boddem, P. Petritsch and D. Mewes, Tomographic imaging of the phase distribution in two phase plug flow, *Int. J. Multiphase Flow*, **24** (1998) 4, 617–634.
7. F. Natterer, The mathematics of computerized tomography, B.G. Teubner-Verlag, Stuttgart, 1986.
8. G.T. Herman, Image reconstruction from projections: The fundamentals of computerized tomography, Academic Press, New York, 1980.
9. A.C. Kak and M. Slaney, Principles of computerized tomography imaging, IEEE Press, New York, 1988.

Computerized Tomography for Scientists and Engineers
Edited by P. Munshi
Anamaya Publishers, New Delhi, India

2. Impedance Technique for the Measurement of Two Phase Flow Parameters: Possibilities and Challenges

P.K. Das[1], G. Das[2], S. Sen[3] and K. Biswas[3]

Department of [1]Mechanical Engineering, [2]Chemical Engineering and [3]Electrical Engineering, Indian Institute of Technology, Kharagpur-721 302, India

Abstract

Measurement of two phase flow parameters poses continuous challenge to scientists and engineers. Impedance technique is versatile and offers some unique advantages for such measurement. In this article, the merits and limitations of the impedance technique have been highlighted. Reference has been made to some of the authors' past works and present research activities.

1. Introduction

Multiphase flow or simultaneous flow of several phases is commonly encountered in a variety of engineering processes. Simultaneous flow of as many as four phases, namely, water, crude oil, gas and sand is not uncommon during oil exploration, though flow of two phase mixtures is the most common occurrence in industry. Simultaneous flow of gas and liquid or gas liquid two phase flow is observed in power generation, refrigeration and cryogenics as well as in chemical process industries.

In spite of the extensive volume of past research activity, two phase flow is not yet an area in which theoretical prediction of flow parameters is generally possible. Indeed, this situation is likely to persist for the foreseeable future. Thus, the role of experiment and parametric measurement is particularly important.

The techniques of measurement for single phase flow are well established. Based on these techniques, various meters and instruments have been developed which are successfully employed for industrial measurement as well as for R&D activities. Unfortunately, these instruments cannot be directly used for multiphase flow measurement. Most of the problems in multiphase flow measurement arise from the fact that the parameters characterizing it are many times larger than those in single phase flows. In single phase flow, the flow regimes encountered are laminar, turbulent and a transition region between them. In multiphase flow, numerous flow regimes are possible. The flow regimes observed under a specific set of flow conditions is dependent on flow geometry (size and shape) and orientation (vertical, horizontal and inclined), flow direction in a vertical or inclined flows (up or down), phase flow rates and properties (density, viscosity, surface tension). Some typical flow regimes in vertical and horizontal gas liquid flow are shown in Figs. 1 and 2.

It may be noted that the flow regimes change drastically from the adiabatic flow when phase transition (due to heat transfer) is involved. In general, the following features, which complicate the flow situation, can be identified:

1. Number of parameters is large.
2. Even for the steady flow rates of the phases at the inlet, the local flow phenomenon may become intermittent.
3. The distribution and velocity of the phases may change both with space and time.
4. In case of phase change, the flow regimes vary substantially along the flow direction.

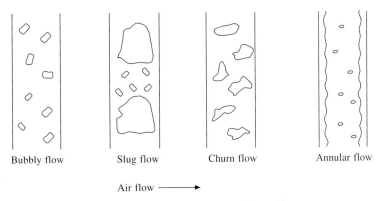

Fig. 1 Flow patterns in vertical up flow.

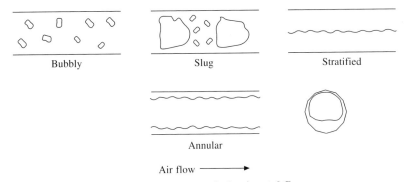

Fig. 2 Flow patterns in horizontal flow.

It may also be appreciated that unlike the single phase flow there are some unique parameters in two phase flow, which need to be measured. For example, as the hydrodynamics depends largely on the prevailing flow regime, it has to be identified and the void fraction or the phase fraction (amount of a particular phase in a given cross section or volume) is to be determined. In certain applications parametric values of each of the phases (namely, phase velocity or phase flow rate) need to be predicated.

This has given rise to the development of a number of techniques especially suited for the measurement of two phase flow parameters. As often one has to measure the parameters separately for both the phases, one needs to exploit a property, which has grossly different values for the phases at the condition of measurement. Based on this logic, measurement techniques have been developed using optical, radiation attenuation acoustic and impedance principle (Hewitt, 1978). This article elaborates the measurement techniques based on impedance principle.

2. Impedance Technique

As the electrical impedance of a two phase mixture is a function of concentration, measurement of impedance can form a basis for the estimation of two different phases. Several instruments for the measurement of void fraction and associated parameters have been developed based on impedance technique. This technique has the following advantages:

1. It is a low cost technique.
2. It is suitable for transient measurement.
3. Large variations of electrode design are possible making the method appropriate for different flow situations and geometry.
4. Both intrusive and non-intrusive measurements are possible.
5. This does not pose any health hazard.
6. Point measurement as well as global measurement can be made by suitable design of the probe.
7. Same principle (sometimes the same probe) may be used for the measurement of associated parameters, viz. (a) flow regime identification, (b) bubble size and frequency and (c) bubble velocity.

An impedance probe can be designed such that the prediction is made due to the variation of either capacitance or resistance. If the liquid phase is the continuous one and electrically conducting, then the probe is used in the resistive mode. If the gas phase is continuous or the liquid phase is non-conducting, then the probe is used in capacitance mode. A large variety of probe designs are possible. Some of them are described in later sections.

Impedance measurement can also be used for assessing some very important qualitative trends of two phase flow like flow regime. Fig. 3 (Das et al., 1999) shows different flow regimes identified with the help of a conductivity probe.

3. Measurement of Void Fraction

Impedance technique is extensively used in determining void fractions. Basically two different principles used are: (1) averaging, i.e. point average, chordal line average, area average and volume average; and (2) tomographic.

3.1 Point Measurement

This measurement principle invariably exploits the difference in electrical conductivities of the two phases to detect the presence of a particular phase at a small area or 'point'. It is extensively used when the continuous phase is a conducting one and the dispersed phase is non-conducting, e.g. gas-liquid bubbly flow. The phase has two electrodes, one in the form of a needle with only its tip tree from insulation and the other with large area of exposure. Both are dipped in the two phase mixture. As the dispersed phase (say bubble) contacts the needle tip there is an abrupt change in the signal level. Ideally, the probe should give binary valued signal. Fig. 4 shows a needle contact probe.

The time trace of the signal should be as given in Fig. 5(a). The square peaks and valleys denote the presence or absence of a particular phase at the probe tip. Thus, the local void fraction (time averaged) can be determined from the following:

$$\alpha_{point} = \frac{\Sigma t_{gas}}{\Sigma t_{gas} + \Sigma t_{liquid}}$$

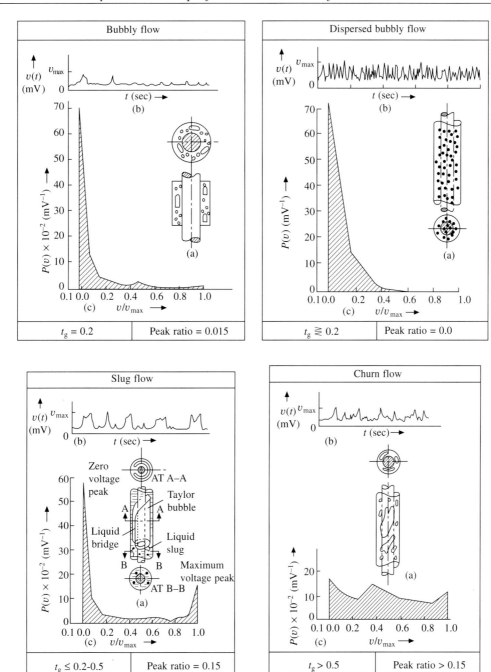

Fig. 3 Different flow regimes detected by parallel plate probe.

However, in actual practice the probe response due to change of phase is not instantaneous. Bubble deformation, wetting of probe and film drainage distorts the signal. There may also be a shift of the base line. The actual signal is shown in Fig. 5(b).

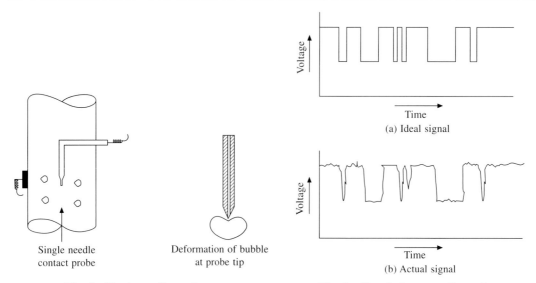

Fig. 4 Single needle probe.   Fig. 5 Signals from needle probe.

A single needle probe may be used with traversing arrangement for the estimation of void fraction in a cross section. Fig. 6 shows the experimentally determined voidage profile (Murali, 1993).

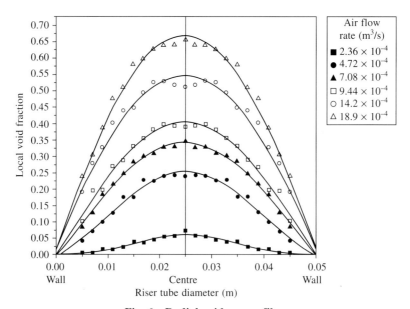

Fig. 6 Radial voidage profile.

Double needle probes may be used for the determination of the bubble velocity, while probes with three or four needle tips may be used for determining different geometrical parameters of a single bubble. These two probes are shown in Fig. 7.

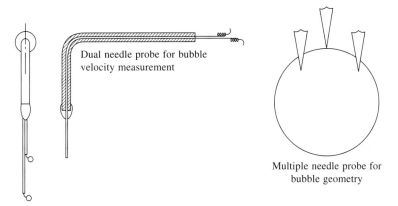

Fig. 7   **Multiple needle contact probes.**

The method of the void fraction measurement is rather straightforward and does not need any mathematical analysis. However, calibration may be required. Besides, the probe is intrusive and may distort the flow phenomena.

There are other types of local probes. Fig. 8 shows different probe geometries for local film thickness measurement.

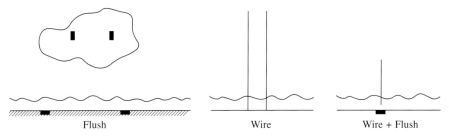

Fig. 8   **Probes for local film thickness.**

3.2 Area Based Measurement

The electrodes have finite areas and the response depends on the concentration and distribution of the two phases in the cross sectional area circumscribed by two electrodes. Arc electrodes shown in Fig. 9 are suitable for circular tubes.

Both conductance and capacitance probe can have such electrode design. We may estimate the void fraction using mathematical formulation. These probes are suitable for simulations (i) when the two phases are completely separated as in case of stratified flow or annular flow; or (ii) where the two phases are homogeneously mixed like bubble or droplet flow. The analysis methods for these two cases are discussed as follows.

(i) Separated Flow of Gas and Liquid

In annular flow, stratified flow, film flow, the gas and liquid phases are separated by a well defined interface and generally the liquid phase does not contain any gas bubble in dispersed condition. In such a situation, impedance probe gives a good estimate of the liquid film thickness. A case of stratified flow is illustrated in Fig. 9.

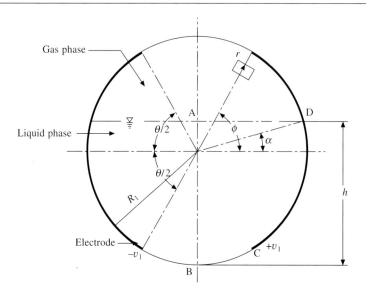

Fig. 9 Schematic representation of the stratified gas-liquid system—arc electrode probe.

From electrostatics:

$$\nabla^2 V = 0 \tag{1}$$

$$E = -\nabla V E \tag{2}$$

$$J = \varepsilon E \tag{3}$$

From Ohm's law

$$J = \iint_s J ds \tag{4}$$

and

$$R = 2V_1/I \tag{5}$$

where V is the electrical potential, J the current density, E the electrical field, I the current, ε the conductivity of the liquid and R the resistance.

Finally, the current I can be expressed as

$$I = L\sigma \int_{-\theta/2}^{\theta/2} E_r(R_1 d\Phi) \tag{6}$$

From the above relationships, we can calculate the resistance for a given liquid height and liquid conductivity as

$$R = |V_{in}/V_0| R_f \tag{7}$$

Using similar relationships one can determine the void fraction for different separated flow conditions.

(ii) Homogeneous Mixed Flow of Gas and Liquid

Impedance probes can also be used when the phases are well mixed (bubbly flow or drop flow). When the phases are well mixed, from Maxwell's theory, we get

$$\alpha = \frac{A - A_c}{A + 2A_c} \cdot \frac{\varepsilon_G + 2\varepsilon_L}{\varepsilon_G + \varepsilon_L} \qquad (8)$$

where A_c is the admittance of the gauge when immersed in the liquid phase alone, ε_G and ε_L are, respectively, the conductivities of the gas and liquid phases if conductivity is dominating. On the other hand, we should use the dielectric constants if capacity is important. Eq. (8) is suitable for bubbly flow.

For liquid droplet flow through a gas, we get

$$\alpha = 1 - \frac{(A\varepsilon_L - A_c\varepsilon_G)}{(A\varepsilon_L + 2A_c\varepsilon_G)} \cdot \frac{(\varepsilon_L + 2\varepsilon_G)}{(\varepsilon_L - \varepsilon_G)} \qquad (9)$$

Arc electrode probes are flush mounted on the wall; hence they are non-intrusive. However, there is fringing effect at the two electrodes. This may be reduced by using guard electrode as shown in Fig. 10.

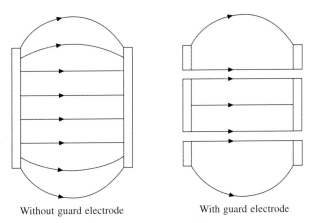

Without guard electrode   With guard electrode

Fig. 10 Fringing effects.

Gupta et al. (1994) used this scheme for estimating the void fraction in stratified two phase system as shown in Fig. 11.

Parallel wire probes have some unique features. Though they are intrusive, the area blockage produced by them is very small. Such probes have been used for estimating the void fraction in slug flow, stratified flow as well as annular flow. Fig. 12 shows a parallel wire probe.

If the interphase position changes widely in a cross section, a number of parallel wires may be used as shown in Fig. 13.

3.3 Volume Based Measurement

All the measurements described above are volume based. Any electrode design will have its corresponding sampling volume. However, by suitable precaution we may approximate the measurement as point- or area-based. On the contrary, the probes discussed in the present section are deliberately designed to sense the two phase mixture over a finite volume. Fig. 14 shows ring

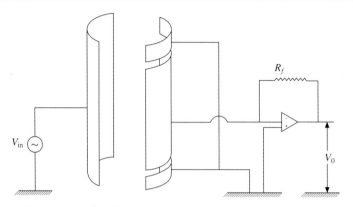

Probes and measuring circuit = | V_{in}/V_0 | R_f

Fig. 11 Measuring circuit with guard electrodes.

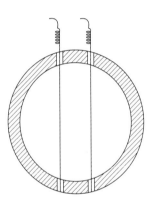

Fig. 12 Parallel wire probe.

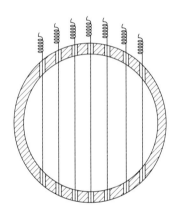

Fig. 13 Multiple parallel wire probe.

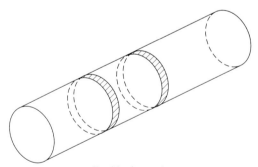

Double ring probe

Fig. 14 Ring electrode probe.

type electrodes, which are mounted on the pipe wall non-intrusively, for sensing the region between the two rings.

Fig. 15 shows a pair of grid electrodes, which are used to measure the void fraction of the two phase mixture in between them. Volume based measurements can be made both intrusively and non-intrusively.

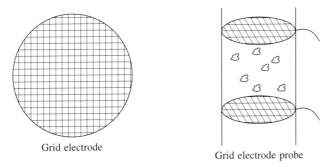

Fig. 15 Grid electrode probe for volumetric measurement.

There can be different configurations of capacitance probes also (Sami et al., 1980) as shown in Fig. 16.

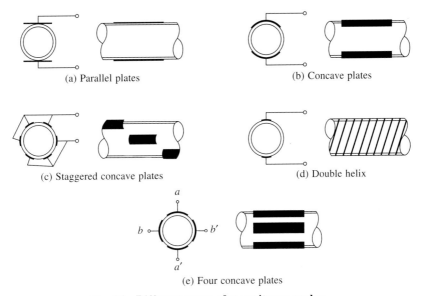

Fig. 16 Different types of capacitance probes.

4. An Integrated System for Measuring Mass Flow Rate in a Pneumatic Conveying Line

In this section a mass flow meter for solid gas two phase system is described. The meter is based on inference technique where the solid velocity and concentration are determined separately and the mass flow rate is inferred from these two information.

Velocity is estimated by cross correlating the signals obtained from two electrodynamic probes. When solid particles move through conduits they generate static electricity due to friction between the particles and friction between particles and the wall. This signal is picked up by electrodynamic probe. Solid concentration is measured by a capacitance probe. On-line data acquisition has been done with the help of a PC. The details of the work is described by Sen et al. (2000). Fig. 17 describes the experimental set-up. Fig. 18 shows the signals from the electrodynamic

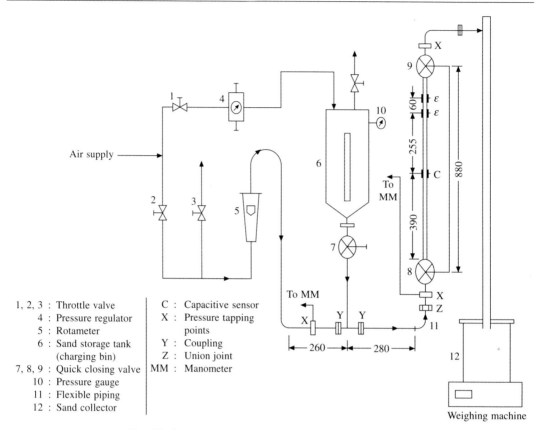

Fig. 17 Integrated system for mass flow measurement.

1, 2, 3 : Throttle valve
4 : Pressure regulator
5 : Rotameter
6 : Sand storage tank (charging bin)
7, 8, 9 : Quick closing valve
10 : Pressure gauge
11 : Flexible piping
12 : Sand collector

C : Capacitive sensor
X : Pressure tapping points
Y : Coupling
Z : Union joint
MM : Manometer

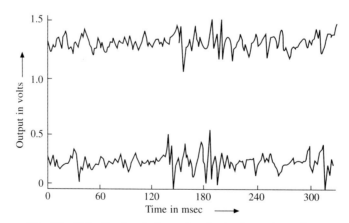

Fig. 18 Velocity signals from the electrodynamic probes.

probes while their cross correlation function are shown in Fig. 19. Finally, results of the mass flow measurement are shown in Fig. 20.

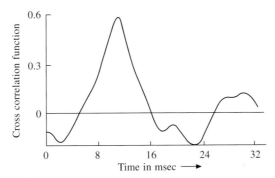

Fig. 19 Cross correlations of the velocity signals.

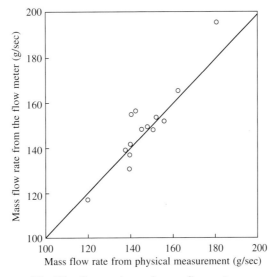

Fig. 20 Comparison of mass flow rates.

5. Electrochemical Behavior of the Electrode/Solution Interface

When a metal electrode (as used in conductivity probe) comes in contact with a solution, the electrochemical processes which may occur at the electrode/solution interface are:

(a) Oxidation-reduction process between the solution and the electrode.
(b) Formation of double layer capacitance.
(c) Faradaic reaction at the electrode.
(d) Diffusion charge transfer to and from the electrode.

(a) Oxidation-reduction process between the solution and the electrode

The simplest chemical process at an electrode/solution interface can be described by this process. This is represented by an ohmic resistance R_{sol} (Fig. 21). But in practice the reactions at the electrode surface are often complex and may include additional processes (McDonald, 1987; McNaughtan et al., 1999).

Fig. 21 Resistance due to oxidation-reduction.

(b) Formation of double layer capacitance

The double electric or dipole layer is formed owing to the action of the applied electric field, the difference in chemical potential between the metal ions of the electrodes and the ions in the solution, and the specific adsorption of ions and polar molecules (Preobranosky, 1980).

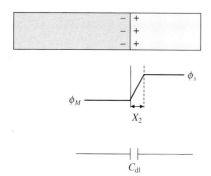

Fig. 22 Helmholtz model.

The double layer behavior is described by various models. One of the simplest models is the Helmholtz model (Fig. 22) in which the excess charge on the solution side simply collects in a thin slab near the metal surface. We have a charge separation and hence a simple capacitor called the *double layer capacitance* (C_{dl}).

This model neglects the fact that the conductivities of the two sides of the capacitor are distinctly different, requiring a thicker layer of charge on the solution side of this double layer, so-called *diffuse layer of charge*. This clearly indicates that a more advanced treatment of the charge interface is required.

Gouy-Chapman treated with diffuse part of the double layer and this model along with the Helmholtz model together proposed by Stern gives the well known *Gouy-Chapman-Stern* (GCS) model of the electrode/electrolyte interface.

The Gouy-Chapman-Stern theory demonstrated mathematically that the double layer behaves as two capacitors in series. One as *inner layer capacitance* (C_i) and the other *diffuse layer capacitance* (C_d) so that $1/C_{dl} = 1/C_i + 1/C_d$ (Bard et al., 1980).

When water is surrounding the electrode then the Gouy-Chapman-Stern theory is also inadequate. In the Helmholtz model of the interface, the capacitance of the electrode/electrolyte interface was considered constant with changes in potential, and using this simple approach, the dielectric constant was determined to be roughly 10, for X_2 approximately the diameter of a water molecule. The normal bulk dielectric for water is 78.5, which illustrates the inadequacy of the simple model and also brings up a point that it may be inappropriate to use bulk values to describe molecular properties. This is specially true for water, which has a strong dipole and will be influenced by the charge on the metal surface (Bockris and Reddy, 1970). GCS model (Fig. 23) treated the solvent as a continuum and did not explicitly include the effect of the displaced ions. In a model proposed by Bockris, Devanthan and Muller, the displacement of water molecule by

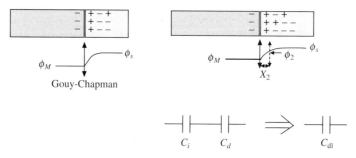

Fig. 23 Gouy-Chapman-Stern model.

specifically adsorbed ions (usually anions) has been considered. According to them, the first layer of water, which is held tightly to the metal, is thought to have a dielectric constant of 6. The second layer of water, which is less tightly held is more free to rotate having a dielectric constant of 30-40.

In summation we can say that the formation of double layer is a complicated phenomenon. The value of the double layer capacitance is dependent on the potential applied to the electrode, electrolyte concentration and nature of the solution. The double layer capacitance is also dependent on the area of the electrode. This capacitance is independent of supply frequency (Preobranosky, 1980). Typical values for the double layer capacitance are in the region of 10-100 $\mu F/cm^2$ (McNaughtan et al., 1999).

Sometimes the double layer capacitance is replaced by *constant phase element* (CPE) because in electro impedance spectroscopy it has been found that the centre of the capacitive loop deviates from the real axis, and the loop diverges from a true semicircle. This deviation originates from the frequency dispersion of the capacitance by dielectric relaxation, and is interpreted by an equivalent circuit involving constant phase element (CPE) (Zoltowski, 1998; Itagaki et al., 2002).

(c) Faradaic reaction at the electrode
If a Faradaic reaction occurs at the electrode, a Faradaic impedance parallel to the double layer will be found. In case of a simple irreversible reaction, this Faradaic impedance is a pure resistance and is called *charge transfer resistance*. Thus, the possible equivalent circuit models for the interface layer can take the form as shown in Fig. 24.

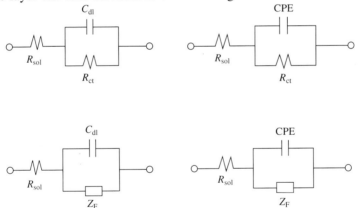

Fig. 24 Different circuit models to describe electrode/solution interface (C_{dl} : Double layer capacitance, Z_F : Faradaic impedance, R_{ct} : Charge transfer resistance, CPE : Constant phase element and R_{sol} : Resistance of the solution).

(d) Diffusion charge transfer to and from the electrode
In case a charge transfer is also influenced by diffusion to and from the electrode, the Warburg impedance is introduced in the circuit (Fig. 25). The complete circuit is often called *Randles' circuit* for electrochemical cell.

It is evident from the above discussion that the electrochemical interaction at the electrode-solution interface is a complex phenomenon. From measurement point of view it is very difficult to characterize the interaction as sometimes the reaction is unpredictable. In electrochemistry the impedance is measured by spectroscopy, but the assumptions are very difficult to meet.

One way to overcome the problem is to insulate the probe to minimize the interaction between the metal surface of the electrode with the solution and measure in capacitive mode.

The advantage of measuring in capacitive mode is that the response is linear with the length of probe dipped inside water. Moreover, the measurement is more consistent, as any insulation layer deposited on the electrode does not hamper the measurement (Baxter, 1997).

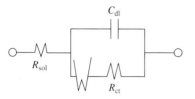

Fig. 25 **Randles' circuit for electrochemical cell.**

From the above discussion it is clear that electrochemical phenomena at the probe solution interface is very complex and difficult to model. We have taken different strategies to reduce it. Some of the results are described as follows.

Table 1 shows the comparison of the capacitance values between an uninsulated and insulated probe, measured with a precision LCR meter (HP4284A).

The probes are made of 6 mm wide strip cut-out from a double sided copper cladded printed circuit board. The thickness of the dielectric material (epoxy glass) between the copper electrodes is 1.62 mm and length 12 cm.

Two different insulation materials used are (a) paper used for insulating purpose of transformer coil and (b) the mixture of resin and hardner often used for adhesive.

Measurements are taken dipping the probe in the water. The results are shown both in Table 1 and Fig. 26. The improvement in the performance is obvious when insulation is applied.

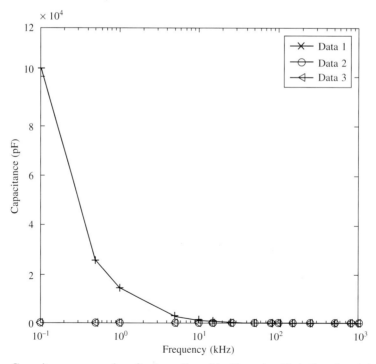

Fig. 26 **Capacitance comparison for 1 cm dipped inside water (Data 1: uninsulated probe, Data 2: probe insulated with paper, Data 3: probe insulated with adhesive).**

Table 1. Comparison of the values of capacitance

Frequency (Hz)	In air (pF)	1 cm dipped in water			5 cm dipped in water		
		Un-insulated probe (pF)	Probe electrode insulated with paper (pF)	Probe insulated with adhesive (pF)	Un-insulated probe (pF)	Probe electrode insulated with paper (pF)	Probe insulated with adhesive (pF)
100	6.72e + 01	1.03e + 05	6.76e + 02	2.87e + 02	3.92e + 05	1.33e + 03	6.44e + 02
500	5.61e + 01	2.55e + 04	3.42e + 02	1.32e + 02	1.04e + 05	6.25e + 02	2.89e + 02
1 k	5.25e + 01	1.41e + 04	2.56e + 02	1.05e + 02	5.72e + 04	4.82e + 02	2.29e + 02
5 k	4.46e + 01	2.81e + 03	1.26e + 02	6.95e + 01	1.16e + 04	2.85e + 02	1.50e + 02
10 k	4.17e + 01	1.28e + 03	1.01e + 02	6.02e + 01	5.31e + 03	2.43e + 02	1.29e + 02
15 k	4.01e + 01	8.09e + 02	9.20e + 01	5.59e + 01	3.28e + 03	2.26e + 02	1.19e + 02
25 k	3.82e + 01	4.45e + 02	8.24e + 01	5.13e + 01	1.74e + 03	2.07e + 02	1.09e + 02
50 k	3.59e + 01	2.05e + 02	7.28e + 01	4.64e + 01	7.41e + 02	1.87e + 02	9.73e + 01
80 k	3.44e + 01	1.29e + 02	6.80e + 01	4.37e + 01	4.26e + 02	1.74e + 02	9.07e + 01
100 k	3.38e + 01	1.07e + 02	6.61e + 01	4.26e + 01	3.34e + 02	1.69e + 02	8.78e + 01
150 k	3.28e + 01	7.97e + 01	6.29e + 01	4.08e + 01	2.23e + 02	1.59e + 02	8.31e + 01
250 k	3.17e + 01	6.05e + 01	5.93e + 01	3.89e + 01	1.47e + 02	1.47e + 02	7.77e + 01
500 k	3.05e + 01	4.88e + 01	5.4e + 01	3.67e + 01	1.01e + 02	1.28e + 02	7.12e + 01
800 k	2.99e + 01	4.39e + 01	5.04e + 01	3.54e + 01	8.72e + 01	1.13e + 02	6.70e + 01
1 M	2.97e + 01	4.27e + 01	4.86e + 01	3.47e + 01	8.31e + 01	1.05e + 02	6.49e + 01

6. Conclusion

In the previous sections different common usage of impedance technique in two phase flow measurement has been discussed. They can be used conveniently in identifying flow regimes and estimating the void fraction in specific situations. However, following shortcomings of the impedance technique have to be kept in mind

(i) Estimation of void fraction: As the probes operate, neither purely in the capacitance mode nor in the resistance mode, theoretical modeling becomes difficult. Also the assumptions made in the theoretical formulation make the result different from actual measurement.

(ii) Fringing effect: This results in three-dimensional effects in area probes and induces error. In conductivity probes this can be reduced by guard electrodes. The size of the capacitance probes needs to be optimized to reduce the fringing effect.

(iii) Stray capacitance effect: As the change of capacitance due to the change of phase fraction is less, the circuit has to be designed properly to reduce the stray capacitance effect.

(iv) Property variation: The change of conductivity and dielectric constants due to impurity, temperature change is also another concern.

(v) Double layer effect: This is one of the least understood phenomena, which plays a very

important role in the performance of the impedance probe. With the existing information it is extremely difficult to model the effect of double layer. However, by suitable design of the probe and adoption of experimental technique this effect can be reduced. Further research should be directed towards this direction.

References

Bard, Allen J. and Faulkner, Larry R. Electrochemical Methods: Fundamentals and Applications, John Wiley & Sons, 1980.

Baxter, C. and Larry, K. Capacitive Sensors Design and Application, IEEE Press, New York, 1997.

Bockris, John O'M. and Reddy, Amulya K.N. Modern Electrochemistry, an introduction to an interdisciplinary area, Vol. 2, Plenum Press, New York, 1970.

Das, G., Das, P.K., Purohit, N.K. and Mitra, A.K. Development of flow pattern during cocurrent gas liquid flow through vertical concentric annulus. Part I: Experimental Investigations, *Trans. ASME, J. Fluids Engg.*, **121**, pp. 895–901, Dec. 1999.

Gupta, D., Sen, S. and Das, P.K. Finite-differences resistance modeling for liquid level measurement in stratified gas-liquid systems, *Measurement Sc. & Tech.*, **5**, pp. 574–579, 1994.

Itagaki, Masayuki, Taya, Akihiro, Watanabe, Kunihiro and Noda, Kazuhiko, Deviation of Capacitive and Inductive Loops in the Electrochemical Impedance of a Dissolving Iron Electrode, *The Japan Society of Analytical Chemistry*, **18**, pp. 641–644, 2002.

Hewitt, G.F. Measurement of Two Phase Flow Parameters, Academic Press Inc., New York, 1978.

McDonald, J.R. Impedance Spectroscopy, Emphasizing Solid Materials and Systems, John Wiley & Sons, New York, 1987.

McNaughtan, Meney, K. and Grieve, B. Electrochemical Issue in Impedance Tomography, 1st World Congress on Industrial Process Tomography, Buxton, April 14–17, pp. 344–347, 1999.

Rao, N.M. PhD Thesis on investigations on buoyancy induced circulation loops, IIT Kharagpur, India, 2003.

Preobranosky, V. Measurement and instrumentation in Heat Engineering, Vol. II, Mir Publishers, Moscow, 1980.

Sami, M., Abouelwafa, A., John, E. and Kendall, M. The use of capacitance sensors for phase percentage determination in multiphase pipelines, Heat Transactions on Instrumentation and Measurement, Vol. IM-29, No. 1, March 1980.

Sen, S., Das, P.K., Dutta P.K., Maity, B., Chaudhury, S., Mandal, C. and Roy, S.K. PC-based gas-solids two-phase mass flowmeter for pneumatically conveying systems, Flow Meaurement and Instrumentation 11, pp. 205–212, 2000.

Zoltowski, Piotr. On the electrical capacitance of interfaces exhibiting constant phase element behavior, *Journal of Electro Analytical Chemistry*, **443**, pp. 149–154, 1998.

Computerized Tomography for Scientists and Engineers
Edited by P. Munshi
Anamaya Publishers, New Delhi, India

3. X-Ray Computed Tomography for Solid Objects

A.K. Jena, N.K. Das, S.J. George, B. Venkataraman, C. Babu Rao,
K. Kasiviswanathan, T. Jayakumar, P. Kalayanasundaram and Baldev Raj
Division for PIE and NDT Development, Indira Gandhi Centre for Atomic Research, Kalpakkam-603 102, India

Abstract

A tomography system has been developed at Indira Gandhi Centre for Atomic Research, Kalpakkam. The system consists of a 4-axis manipulator, an X-ray source and a real time detector. The complete automation and data collection software has been developed in-house. Perfect cylindrical objects and cylindrical objects with central and off-centre holes made of aluminum have been scanned using this system. Beam hardening corrections have been carried out. Reconstruction software based on convolution back projection algorithm with filter has been developed and tested. The reconstructed images are close to actual objects.

1. Introduction

Computed tomography (CT) is an imaging technique, which produces cross sectional details of an object from its line integrals of projections. The design of the CT machine for industrial and scientific applications is more complex than medical systems, since it has to accommodate large variation in density and achieve finer resolution. A tomography system has been developed at Indira Gandhi Centre for Atomic Research, Kalpakkam with a 4-axis manipulator, an X-ray source and a real time detector. The X-ray source and detector are mounted on a 'C' arm, which makes the alignment of source and detector simple. The 4-axis precision manipulator has been designed and developed indigenously, which can handle objects upto 25 kg. The maximum diameter of the object can be 20 cm. A few standard test objects like perfect cylinder, cylinder with central and off-centre holes were fabricated in aluminum with a tolerance of 5 micron. The tomograms of these objects were realized using this system. The beam hardening corrections were made using various thicknesses of aluminum blocks. To check the procedure of beam hardening corrections, an aluminum cylinder was fabricated with 3 mm central hole filled with steel. The density variation of the steel is predicted within 7% after beam hardening correction.

For reconstruction of image, the fan beam data were converted to parallel beam and Filtered Convolution Back Projection Technique and Shepp Logan filter was used. The reconstruction program was developed in-house using FORTRAN 77 language. For image display, Visual C++ was used. The reconstructed images are good considering the detector and the data acquisition system used.

2. Set-up

The set-up consists of four main components, viz. manipulator, X-ray source, real time detector and data acquisition PC (Fig. 1). The specifications are as follows:

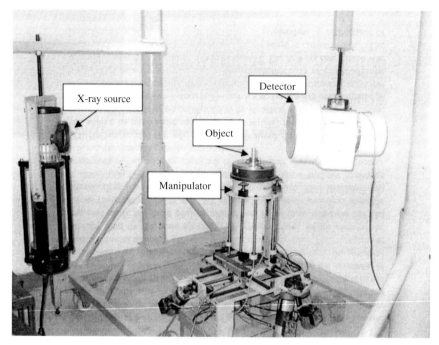

Fig. 1 X-ray tomography set-up.

Manipulator
Span: X 300 mm, Y 250 mm and Z 150 mm

Resolution
Linear: 10 micron, Angular 0.001°, Maximum weight: 25 kg
Maximum diameter of object: 200 mm
Motors: Servo
Controller: GALIL Motion Control System

X-ray Source
Beam: Directional
Voltage range: 10-200 kV
Current: 12 mA maximum
Maximum power: 900 W
Focal spot size: 1.5 mm
Cooling: Air cooled

Real Time Detector
Screen: CsI (Tl) 100 micron
Resolution: 45 lp/cm (max)
Magnification: 1.0, 1.5 and 2.0
Multiplication: 14,000
Camera: Analogue CCD

Data Acquisition PC
Intel PII 333 MHz processor with 2 GB hard disk
FALCON frame grabber card

3. Data Collection

In radiography, shadow of the three dimensional object is produced by irradiating the object with X-rays. In tomography, the cross sectional details are estimated from the projections (or simply ray integrals) of the transmitted rays obtained at different angles. The projections are also known as *radon transforms* of the image, named after their originator [1]. The algorithm developed in the beginning required a great effort to reduce computing time. Present day personal computers can reconstruct the image in less than 10 sec with reasonable resolution and accuracy.

The schematic diagram of the system used for data collection (Fig. 2) shows that the X-ray source is independent of the automated data collection system. The manipulator and the real time detector systems are coupled by means of digital I/O card. For scanning these objects, 0.5 mA current at 58 kV potential was used. One typical image seen in the real time system using FALCON [2] image acquisition card is shown in Fig. 3. This is a 25 mm cylindrical object with 3 mm steel rod at the centre. The projected tip used for obtaining centre pixel data is machined to a diameter of 1 mm. The data for each object was acquired in steps of 0.9 degree. These were in 8 bit resolution. The data acquisition for each object takes around 18 min.

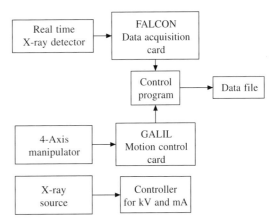

Fig. 2 Schematic diagram of the system for data collection.

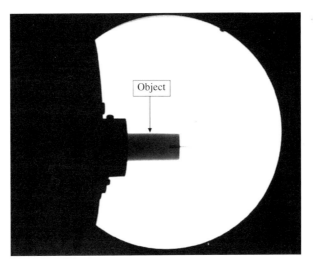

Fig. 3 Image acquired by FALCON frame grabber card.

For beam hardening correction [3], centre pixel data were collected for various thicknesses of aluminum slabs. The procedure followed for beam hardening correction is same as that described in [4]. The attenuation was plotted against thickness as shown in Fig. 4. The line close to x-axis is the observed attenuation for the polychromatic source used. This curve is obtained by fitting the experimental data by a third degree polynomial. The slope of this curve at $x = 0.0$ was determined. The upper straight line was determined using this slope and zero intercept. The measured attenuation was corrected using this curve.

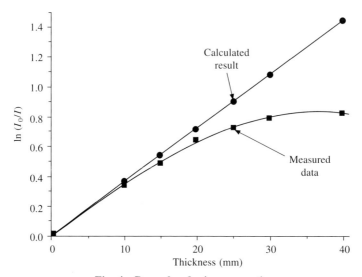

Fig. 4 Beam hardening correction.

4. Reconstruction

The source used in our set-up has fan beam geometry with equispaced detector pixel. The algorithm used for reconstruction is Filtered Back Projection. This algorithm is currently used in all the applications of straight ray tomography. This algorithm has been found to be extremely accurate and amenable to fast implementation using computers [5, 6].

The derivation and implementation details of Filtered Back Projection technique for three types of scanning geometries, parallel beam, equiangular fan beam and equispaced fan beam are dealt in various publications and text books [1, 6, 7]. The computer program is written in FORTRAN language. The procedure followed for reconstruction is as follows:

1. *Conversion of the fan beam to parallel beam:* The experimental configuration and the real time detector used fall in the category of equispaced collinear detectors. Hence, we have followed the procedures that were described in section 3.4.2 of [7]. In order to obtain equispaced parallel beam, data interpolation is required. Linear interpolation was used.

2. *Parallel beam projections were obtained:* A single projection with error margin is shown in Fig. 5. Stacking all the projections together results in a 2D data set $P_\theta(r)$, which is called *sinogram* (Fig. 6). Sinograms for all objects were generated. These give the health of the data acquired. The sinogram for the cylinder with one central hole and two off-centre holes is shown in Fig. 7.

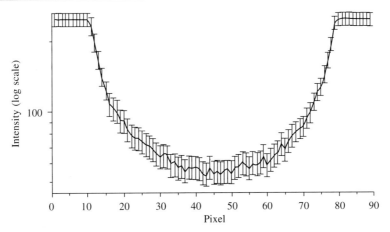

Fig. 5 Projection data for 25 mm aluminum cylinder.

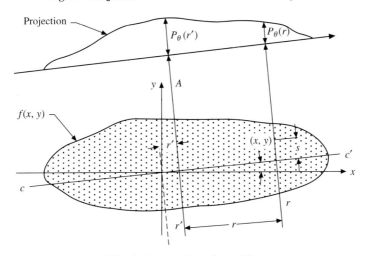

Fig. 6 Projection of an object.

3. *Fourier transforms of the projections are taken:* Standard FFT routine available is used after padding the projection data with zeros to make 128 pixels.

4. *Before applying the inverse FFT, filtering is essential:* Otherwise blurring in the reconstructed image is observed [7]. Shepp Logan filter was generated and multiplied with the above result. This filter is chosen since 8 bit data has more noise.

5. Results are back projected by taking inverse FFT and followed by normalization.

6. Results were displayed using Visual C++ tool.

5. Results and Discussions

The reconstructed image for an aluminum cylinder of 25 mm diameter is shown in Fig. 8. The image does not have sharp boundaries. This is attributed to the data collection using 8 bit ADC. If we use 12 bit or more ADC with high dynamic range detectors, then this problem will be solved. Fig. 9 shows the reconstructed image of 25 mm cylinder with central hole of 2 mm diameter. As expected there is a spread near the boundary. Fig. 10 shows the image of 25 mm

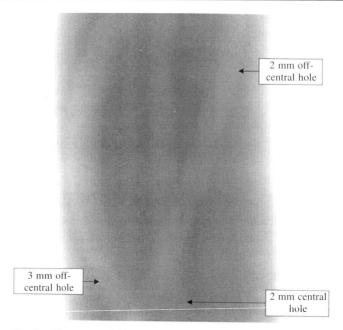

Fig. 7 Sinogram of 25 mm cylinder with 2 mm central hole and
2 mm and 3 mm off-centre holes.

cylinder with one 2 mm central hole and two off-centre holes (2 mm and 3 mm). All these images were obtained after incorporating beam hardening corrections, as described in [4]. The contrast sensitivity for the central hole is about 9%. The delectability of the hole at the centre is 2 mm in the diameter of 25 mm in aluminum. Although theoretically it is feasible to detect 0.5 mm defect/hole at the centre we were not able to detect even 1 mm hole at the centre of 25 mm object. This is attributed to 8 bit data acquisition system. The density variation after beam hardening correction is 7%.

Fig. 8 Reconstructed image of a 25 mm aluminum cylinder.

Fig. 9 Aluminum cylinder with 2 mm central hole.

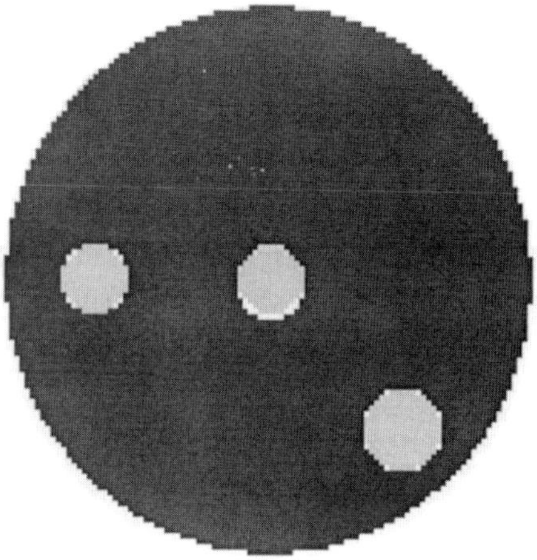

Fig. 10 Aluminum cylinder with 2 mm central hole and 2 mm and 3 mm off-central holes.

As seen from all the reconstructed images, the sharpness is lacking at boundaries and interfaces. This is attributed to the low dynamic range of the detector and the 8 bit resolution of ADC.

6. Conclusions

We have demonstrated that this system can be used for industrial tomography. We have taken up the installation of cooled CCD set-up in order to get better resolution and higher bit data. The system will be subsequently updated for microtomography.

References

1. G.T. Harman, Image Reconstruction from Projections: The Fundamentals of Computed Tomography, Academic Press, New York, 1980.
2. FALCON/EAGLE, IDS Imaging Development Systems GmbH, Module 400, 2001.
3. G.T. Harman, Correction for Beam Hardening in Computed Tomography, *Phys. Med. Biology*, **24**, No. 1, pp. 81–106, 1979.
4. S.F. Burch, X-ray Computerised Tomography for Quantitative Measurement of Density Variations in Materials, *Insight*, **43**, No. 1, pp. 29–31, 2001.
5. M. Soumekh, Image Reconstruction Techniques in Tomographic Imaging Systems, IEEE Trans. Acoustics, Speech and Signal Processing, Vol. ASSP-34, No. 4, Aug. 1986.
6. F. Natterer, The Mathematics of Computerised Tomography, John Wiley & Sons Ltd., 1986.
7. Avinash C. Kak and Malcolm Slaney, Principles of Computerized Tomography Imaging, IEEE Press, New York, 1999.

Computerized Tomography for Scientists and Engineers
Edited by P. Munshi
Anamaya Publishers, New Delhi, India

4. Steady-State Multi-Phase Flow Measurement Facility at FCRI

M. Suresh, R.V. Rajesh, M. Viswanathan and M.S. Konnur
Fluid Control Research Institute (FCRI), Palghat-678 623, India

Abstract

Multi-phase flow, gas-liquid two-phase flow in particular, is of critical importance to industries in oil and gas sector and nuclear power plants. Measurement of multi-phase flow is not easy and the methods currently in use are particularly not very accurate. Uncertainties of unacceptably high order have been reported when conventional single-phase flow meters such as turbine meters, mass flow meters are attempted for multi-phase flows. Extreme variations in viscosity, density and velocity between the phases occur not only across the pipe but also temporarily along the pipe. Much care has to be taken for properties of the different fluids when metering multi-phase fluids. The past three decades have witnessed considerable progress in development of technologies for measurement of phase-fraction and two- and three-phase flow measurement.

As part of a BRNS-sponsored project, FCRI designed and developed a multi-phase flow-meter using gamma-ray attenuation technique in conjunction with venturi meter. A test facility to evaluate performance of the developed multi-phase flow meter was also set up at FCRI. This article describes the multi-phase hydraulic test facility and the performance/behaviour of the multi-phase flow meter under various test flow-regimes.

1. Introduction

Multi-phase flow-metering has been of enormous interest to the petroleum industry especially at onshore/offshore sites since well monitoring, production allocation, reservoir management from hydrocarbon reservoirs and planning of explorations rely heavily on information regarding flow-rates of oil and gas produced by individual wells in the reservoir. Problems in metering of wet-gas and oil-water-gas media have meant use of complex arrangements such as test-separators to meter the individual phases after separation by conventional single-phase flow meters at the production platforms, especially off-shore where space is already at a premium. This is in addition to the inherent economics such as need to install the heavy and bulky devices.

In chemical process industry, the economic aspects in design of process piping involves studies to evaluate the pressure-drop characteristics and flow of two-phase media especially when processes concerned produce fluids that disintegrate or generate two phases. Processes such as condensate return lines flashing into steam, vapour-feed lines of distillation columns and refrigerant-return lines that must maintain a specific vapour-liquid ratio for efficient operation, etc. are examples where considerations for safety, pressure-loss, optimal energy-savings, etc. involve understanding of two-phase flows. In nuclear power plants, it is important to experimentally investigate conditions such as loss of coolant accidents (LOCA) at the crucial design stages itself in order to validate computer codes for prediction of LOCA related transients in two-phase flow.

Multi-phase flow meters in general fall into two categories, viz. metering which require the use of separation of the full/partial gas/liquid fractions upstream of metering point and metering that do not involve use of separators. In the former, the separators used generate a liquid stream and a relatively liquid free gas stream that are metered using single-phase flow measurement techniques that benefit from commercially proven cost competitive technologies and are easier to operate and maintain but require capital and operational expenditures on test-separators. The latter category comprises metering techniques that require either conditioning the flow such as use of in-line mixer or those without any conditioning device. The techniques in the second category, even though eliminate need for conventional test-separators, require use of complex sensing techniques and associated processing electronics and software to generate information on phase fractions, densities and flow-rates.

As part of the BRNS sponsored project, FCRI designed and developed a multi-phase flow measurement system comprising an in line mixer, a dual energy gamma-ray densito meter and a venturi meter downstream of the mixer. A purpose specific test facility was set up to analyze the behaviour/performance of the flow-meter under various operational conditions.

2. The Experimental Facility

The present multi-phase flow test facility shown in Fig. 1 is designed for testing of two-phase flow of air-water composition and comprises 4″ NB multi-phase test line, a reservoir of 12 m^3 capacity for water storage, various flow elements for air and water lines such as filters, control valves, single phase flow meters, pressure/temperature and the associated transducers, an air-water mixer Tee, reference phase-fraction-metering devices, etc.

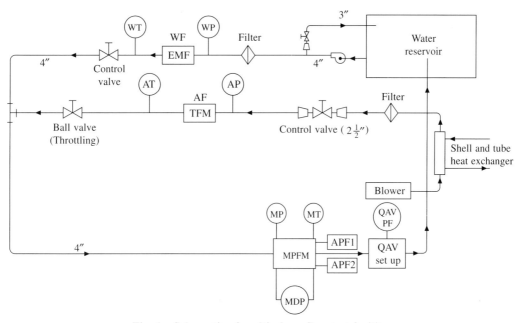

Fig. 1 Schematic of multi-phase flow test facility.

The water loop comprises a 15 HP centrifugal pump, a 50 μm filter, a 4″ control-valve, a 3″ gate valve (on the bypass line), and an electromagnetic flow (EMF) meter. A 60 HP blower-motor assembly delivers air up to 400 m^3/h in the airline at 2.0 bar (abs). The air loop also contains a shell and tube heat-exchanger downstream of the blower, a 5 μm filter, a turbine flow meter (TFM), $2\frac{1}{2}″$ pneumatic-actuated control valve with I/P converter, etc.

The air and water lines are terminated at a mixing Tee wherein the air is injected into the water media to generate the multi-phase (two-phase air water) test-media. The two-phase air-water flow thereafter passes through a straight length of about 150 times the nominal diameter of the pipe to ensure stabilization of flow structure. A Perspex-pipe section at the upstream of the multi-phase flow meter is to visually inspect/record the flow regime for a given air/water velocity combination.

The FCRI developed multi-phase flow meter is mounted at the test-section downstream of the Perspex-pipe. The individual pressure, differential pressure and temperature transmitters (all Smart HART transmitters) and flow meters provide the hydraulic parameters for measurements in the air, water and air-water loops. A reference phase-fraction measurement set-up as shown in Fig. 2 is located downstream of the test-section.

This set-up so designed diverts the flow through either the horizontal or the vertical sections by operation of quick-acting valves (QAV). These QAVs are solenoid-operated pneumatic-actuated quarter-turn (ON/OFF) butterfly valves. During phase-fraction measurements, flow diverted through the vertical (Perspex) section after stabilization is trapped within the vertical section by simultaneous closure of the two vertical QAV. The horizontal bypass section is opened immediately/simultaneously to enable continuation of flow in the test-section. The trapped two-phase media in the vertical section segregates into air and water columns that can be measured to derive the phase-fractions and compare against the gamma-ray phase fraction meter of the developed multi-phase flow meter.

2.1 Multi-phase Flow Meter

The multi-phase flow meter assembly (Fig. 3) comprises an in-line static mixer, a venturi meter and a gamma-ray based phase fraction meter integrated into a vertical-mounting unit.

2.2 Static Mixer

The in-line static mixer is provided in order to obtain a well-established, stable and homogenous mixed-phase media that flows down into the measuring section (at the throat of the venturi) in a form independent of the upstream conditions. This will facilitate the use of multi-phase flow meter for various flow regimes [3] without frequent calibrations/corrections.

The static mixer is a large cylindrical chamber with a horizontal inlet for two-phase media which enables diversion of gas-phase into an upper containment and flowing down of the denser liquid-phase to the bottom as the two-phase media continuously flows into it. The gas from the top chamber gushes out through the injection pipe to the mixing chamber where it mixes with the liquid-phase by turbulent shear mixing process to provide a homogenous media.

The liquid drains from the bottom of the barrel through the venturi, with sparging of gas into liquid accelerating the liquid velocity thereby transferring a part of its energy. The size of the mixer is designed based on expected maximum slug lengths in the test-line. The static mixer is modeled and analyzed through a CFD software (FLUENT) for design optimization for a specified degree of homogeneity in the phase-fraction range.

2.3 Venturi Meter

A venturi meter positioned vertically just below the mixer is used to measure the total mass-flow of the homogenized mixture. For homogenized flow, the standard single-phase flow relations for venturi is applied for mass flow calculations.

Fig. 2   Reference phase-fraction set-up.   Fig. 3   Multi-phase flow meter.

The total mass flow rate for a venturi meter is given by

$$M_t = C_d \frac{A_2}{\sqrt{(1-\beta^4)}} \sqrt{2 \times \mathrm{DP} \times \rho_m}$$

where C_d is coefficient of discharge, $A_2 = \left(\frac{\pi}{4} d_2^2\right)$ is venturi throat area (d_2 is throat diameter), $\frac{1}{\sqrt{(1-\beta^4)}}$ is velocity approach factor $\left(\beta = \frac{d_2}{d_1}\right)$, DP is differential-pressure of homogenized mixture across the venturi and ρ_m the density of the homogenized mixture, obtained from the gamma-ray phase fraction meter.

The venturi is designed as per BS:1042 standard and sizing calculations done using FCRI flow-element sizing-selection software FMSEL. Fabrication of the meter is done in-house at FCRI. The meter is calibrated in the flow laboratories in FCRI separately for air and water media.

2.4   Gamma-ray Phase Fraction Meter

The gamma-ray phase fraction meter installed at the throat of the venturi meter measures the density/phase fraction of the homogeneous air-water media. The unit comprises a gamma-ray emitter (radio isotope source) enclosed in a specially designed source-holder and a gamma-ray detection system.

Gamma-ray Source

Attenuation of gamma-rays occurs to different extents as it passes through different media and it is this property which is made use of in the determination of phase fractions/density of the homogenous phase. Selection of an appropriate source for gamma-ray plays a vital part in the feasibility of phase-fraction measurements. Many factors such as energy-levels, emission-ratio and half-life of the radioactive source, attenuation/losses in pipe-wall, encompassed air-paths and test medium, collimation-distance and beam-size, scintillation-efficiency of the selected detector and measurement/counting time, etc. are very important parameters. At times, use of higher source strength may yield sufficient number of photons and lowers the statistical-error in measurement but leads to safety constraints including storage, shielding and handling difficulties. Moreover, the availability of source indigenously from Board of Radiation and Isotope Technology (BRIT) is also important.

Radio isotope Eu^{152} was selected for the source as it has distinguishable multiple energy-levels (121.78 keV, 344.28 keV, etc.) and good half-life (13.537 years). Statistical measurement errors for photon counting is evaluated on the basis of source strength, energy, detector efficiency, depth and window size of detector crystal, beam diameter, source-detector distance, etc. From statistical point of view, minimum count of 10000 is arrived at for a statistical uncertainty of 1.0% in intensity measurement. The attenuation of gamma-rays is given by

$$N = N_0 e^{-\Sigma \mu_i x_i}$$

where N_0 is the initial disintegrations per second (dps) of the source and N the dps on passing through media i of attenuation coefficient μ_i and thickness x_i.

The source strength S is calculated from the equation

$$N = KS\theta\gamma$$

where $K = \dfrac{\text{Area of collimator hole}}{4\pi l^2}$, l is the distance between source and collimator, S source strength in dps, θ measurement time in sec, γ emission ratio and N number of counts required over the measurement time.

Using the above equations, source strength of about 25 mCi for Eu^{152} is found sufficient for the phase-fraction measurements.

Source-holder

As per mandatory requirements of Atomic Energy Regulatory Board (AERB), radiation sources can be used only in qualified/AERB approved nucleonic gauges that have proper shielding designed to reduce exposure rates to within acceptable levels specified by AERB. Lead is used as shielding material and shielding requirements were calculated based on its half-value layer (HVL) and tenth-value layer (TVL). The exposure rate of europium (Eu^{152}) is 0.53 RHM (Roentgen per hour at 1 metre distance for 1 Ci of source) and design calculations showed that 42 mm thick lead is required to satisfy the safety requirements. The source-holder was therefore designed with 45 mm thick lead present isotropically around the source.

The source-holder after fabrication and assembly was subject to all the required tests specified by AERB. These tests including stray-radiation level test, temperature test, penetration test, drop test, vibration test, endurance test, stacking test, etc., were carried out successfully and the source-holder type/model approved by AERB.

Gamma-ray Detection System

The chief criteria for selection of gamma-ray detector are detection efficiency, scintillator decay constant, and ease of operability in versatile work conditions. The decay constant is important because for a predetermined statistical error, minimum number of photons need to be collected for a preset dwell-time. The decay constant needs to be small to avoid pulse-pile up especially at high-count rates. Higher detection efficiency means reduced source strength and shielding requirements. Widely used scintillators include the inorganic alkali halide crystals (sodium iodide, NaI), and organic-based liquids and plastics. The inorganic crystals are generally more sensitive but slow, while organic scintillators are faster but with lesser light yield. NaI is still the best gamma scintillation detector commercially available, despite of it being hygroscopic [5].

The measurement system comprises an 2″ × 2″ NaI (Tl) scintillation detector with photo-multiplier tube (PMT) base and signal conditioning electronic modules including preamplifier, spectroscopy amplifier, two single channel analyzers (SCA), high voltage supply for PMT, low voltage supply for preamplifier, etc. The block schematic of the system is given in Fig. 4.

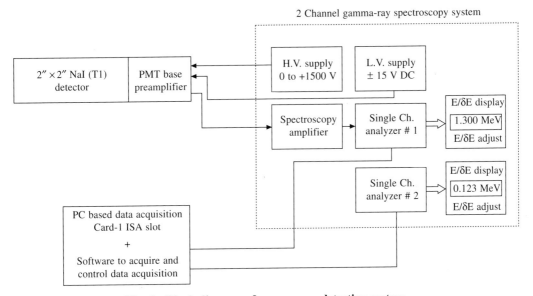

Fig. 4 Block diagram of gamma-ray detection system.

The rack-mount system has individual digital displays for each SCA channels to enable independent indications for discrimination/windowing of individual energy levels. Two separate multi-channel analyzers, ORTEC Trump-PCI-32 (8K MCA) and Para Scientific (2K MCA), were used for detector positioning, ambient/Compton scattering analysis, detector calibration and data crosscheck purposes.

2.5 Instrumentation and Data Acquisition

The instrumentation for the test facility comprises smart pressure, differential pressure (druck/ rosemount, etc.) and temperature transmitters (Terwin), turbine (Hoffer) and electromagnetic (Krohne) flow meters, flow control valves (ILP), quick acting butterfly valves, linear variable differential transducer (Lucas Control Systems), precision rotary measurement system (SONY/ Digiruler), solenoid valves (FESTO), voltage-to-current (V/I) converter, current-to-pressure

(I/P) converter (IL), frequency-to-voltage (F/V) converter (Ectron), 24 V DC power supplies (Phoenix Contact QUINT/PS), triple-output bipolar power supplies, signal-conditioners, limit-switches, etc. besides the electronic modules for the detector system. A PC based data acquisition system (DAS) was set-up with a PCI plug-in multi-functional analog input card (National Instruments/PCI-MIO-16E-1), a 4-channel 32-bit timer/counter PCI add-on card (National Instruments NI-6601) and the MCA (TRUMP-PCI 32), and associated termination boards, external reed-relay interface card (Advantech PCLD-885), etc.

Signals (of type 4 to 20 mA) from different pressure/temperature transmitters and the electromagnetic flow meters, voltage signal (0-5 V) from F/V converter, TTL pulse outputs from the SCA units, etc. are input to the DAS system. Output signals include analog voltage (0-5 V) signal-to-V/I converter to drive the air-line control valve (through the I/P converter), trigger-cum-gating signal (TTL) from counter (NI-6601) to SCA-pulse input gate channels and analog card (NI PCI-MIO-16E-1), digital actuation signals through PCLD-885 to various ON/OFF and QAV valves, etc.

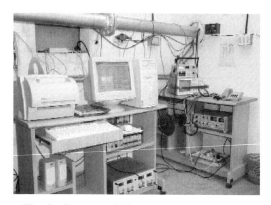

Fig. 5 Data acquisition and control system.

The entire data acquisition program was developed in NI LabVIEW-6i running under Windows'98SE. Fig. 5 shows the data acquisition and control systems connected to a PC. During each run, the system actuates the valves, acquires data from the single-phase and two-phase metering lines including from the SCA channels. The analog measurements are done synchronously with the SCA counting so that the best possible simultaneity is obtained for the homogeneous phase measurements at the multi-phase flow meter. The need for simultaneity of measurements from analog and SCA inputs is critical when measurements are made under conditions that cause small bubbles/slug-flow/phase-slips at venturi.

Towards the end of every acquisition cycle, the QAV system is operated by actuation of QAVs. Once the volume inside the vertical column settles and stabilizes, the venting valve is operated and the stabilized water-column height is measured as differential pressure using a low differential-pressure transmitter.

3. Results and Discussion

3.1 Efficiency of Gamma-ray Detection System

The energy resolution of the NaI based gamma-ray detection system, which utilizes pulse height spectral analysis, is established using a standard reference source of Cs-137 (20 microCurie). The resolution was found to be 6.5% of Cs137. The resolution of 122 and 344 keV are also established and are found to be 13 and 14%, respectively, which will be used as intensity measurement of the specified energy levels. Fig. 6 shows the MCA spectrum of Eu, wherein peaks indicate the gamma energy levels. Graded shielding was used for detector with lead, cadmium and copper sheets.

3.2 Static Calibration Using Lucite Blocks

Static calibration of the phase-fraction meter was done using lucite blocks of known geometry.

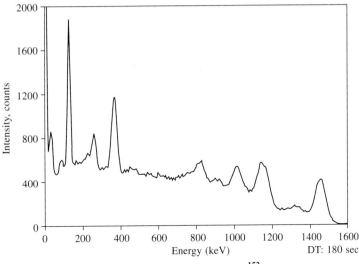

Fig. 6 MCA spectrum of Eu^{152}.

Lucite has density and attenuation properties close to that of water. Lucite blocks were specifically fabricated for simulating full-water, stratified, annular and bubbly flow structures. Fig. 7 shows that the lucite fraction measurements using gamma-ray fraction meter fairly matches those calculated from known lucite-thickness. It was found that the fractions through the gamma-ray fraction meter could be measured with an accuracy of 1.0% for a dwell time of 5.0 sec.

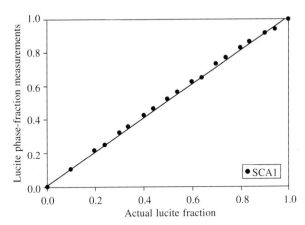

Fig. 7 Results for static calibration using lucite blocks.

4. Multi-phase Air-water Experiments

4.1 Flow Regime Mapping

Flow regime mapping was carried out to identify the structure of flow prevailing in the upstream of the multi-phase flow meter for various combinations of superficial velocities of air and water. In horizontal multi-phase flows, there is asymmetric distribution of phases within the flow channel due to gravitational force acting normal to flow. The heavier fluid tends to accumulate to bottom part of pipe resulting in various flow structures such as bubbly, stratified, wavy, plug, semi-slug, slug and annular flows with respect to superficial velocities of air and water. These flow regimes were observed/recorded on the transparent Perspex-pipe horizontal section upstream of the mixer.

4.2 Differential Pressure Measurements for Optimizing Dwell-time

Though steady input flows are imposed at the upstream of the multi-phase test section, the nature of multi-phase flow is such that there are inevitable fluctuations in venturi inlet pressure and pressure drop which, to a large extent, are dampened by the homogenizer/static-mixer upstream of the venturi meter. Also the preset time for intensity measurement of the gamma-ray phase fraction meter defines the statistical error for measurements. To compute this counting interval, the frequency distributions for differential pressure at the venturi was taken for the entire test protocol to identify the least peak frequency for differential pressure fluctuations. Based on this (about 0.22 Hz) the pulse counting and measurement interval was decided (about 5.0 sec).

4.3 Chordal and Radial Distribution of Phases

The source-detector assembly, gamma-ray phase fraction meter (GRFM), is mounted on a linear sliding mechanism on a rotary table at the venturi meter throat. Phase-fraction and other measurements were taken for every run with the GRFM positioned at seven different pre-determined chordal locations by sliding the linear table. Linear motion is recorded from an LVDT attached to the slide table. The chordal line-phase fractions (Fig. 8) measured across the venturi throat section are then transformed into area averaged phase-fraction through chordal segment inversion (CSI) for comparison with homogenous phase-fraction [1]. Radial symmetricity

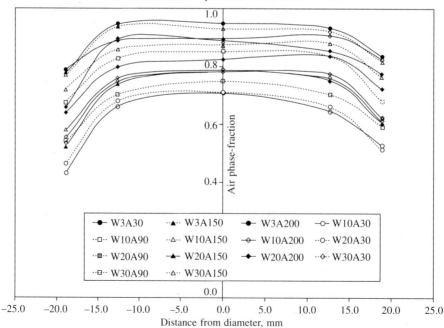

Fig. 8 Chordal distribution.

is assumed to exist at the venturi throat section, which is true in case of downward multi-phase flows and also experimentally verified.

Measurements for determining radial distribution of phases were obtained from the GRFM by rotating the rotary table (GRFM) through various angles. During the experiments, rotational steps of 30 degrees were used. A digiruler magnetic scale was pasted on the rotary table to measure the angular positions. It was observed that the phase-fraction distribution across the venturi throat (Fig. 9) is fairly the same as in the case of homogenous flows.

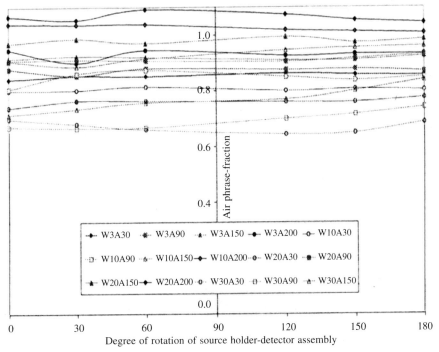

Fig. 9 Radial distribution of phases in venturi throat—SCA2.

4.4 Tomographic Reconstruction

The chordal average method is an estimate of the cross-sectional average whenever the distribution of attenuation coefficient μ (or density ρ) is uniform. For flows with non-uniform distribution, results for estimation of cross-sectional average due to single beam methods become unrealistic. Single beam method is inappropriate in providing information regarding point-wise distribution. Gamma densitometry tomography (GDT) reconstruction develops from the measurement of known intensity along a ray (straight line) passing through an attenuating medium. The intensity of monoenergetic radiation beam I is measured when the beam passes through the medium and is given by Beer Alemberts law

$$\frac{I}{I_0} = \exp\left[-\int \mu\, ds\right]$$

where I_0 is the incident radiation intensity, μ the attenuation coefficient and s the distance along the portion of the ray intersecting the attenuating medium. Attenuation A is given by

$$A = -\ln\left(\frac{I}{I_0}\right) = \int \mu \, ds$$

This shows a linear relation between attenuation A and attenuation coefficient μ. Also, the tomographical reconstruction of the spatial variation of μ can be done if the spatial variation of A is known. Thus, the task of the algorithms is to determine the spatial variation of the attenuation measurements, which are line integrals of μ. The reconstruction is performed using Abel transform [9].

The ray averaged gas hold-up ψ_i on a set of gamma rays is determined according to the relation

$$\psi_i = \frac{\ln(I_{TP}/I_W)}{\ln(I_A/I_W)}$$

where I_{TP} is the intensity of radiation in two-phase flow and I_W and I_A are intensities of radiation when the test section is filled with water and air, respectively. Fig. 10 gives the comparison of gas hold-up values (for a particular flow rate) for various non-dimensional chordal distances obtained by the above equation and through the method of tomographic reconstruction.

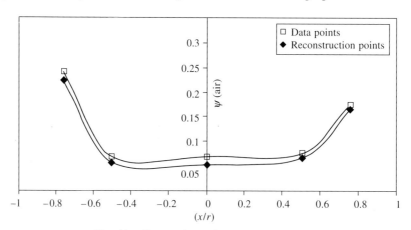

Fig. 10 Comparison of gas hold-up values.

4.5 Total Mass Flow Rate Measurements from Venturi Meter and Gamma-ray Phase-fraction Meter

The line phase-fraction measured at the throat is used for homogenous mixture density estimation. From the mixture density and from the differential pressure of the homogenous mixture across the venturi, total mass flow rate of the mixture is determined. The phase-fractions measured through gamma-ray fraction meters are compared with the homogenous phase-fractions obtained from upstream flow measurements and also with the volume averaged phase-fraction measured through quick closing valves method.

Fig. 11 shows the comparison of the air phase-fractions determined using the GRFM, SCA1 (122 keV) and SCA2 (344 keV). The air phase-fractions estimated through GRFM fairly matches those measured by the QAV method. It is also observed that phase-fractions in the venturi section is less by 10-12% when compared with homogenous phase-fraction as obtained from the individual

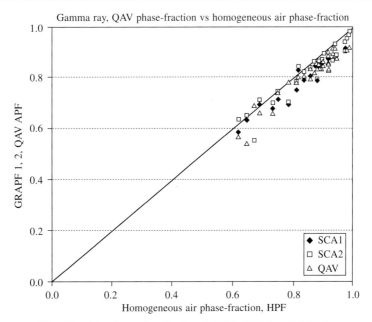

Fig. 11 Air phase-fraction by GRFM, SCA1 and SCA2.

upstream single-phase flow meters. This could be attributed to the degree of homogeneity achieved through static mixer. Fig. 12 shows the comparison of flow rate of water measured from the reference EMF meter and from the multi-phase meter while Fig. 13 compares the air flow rate obtained from the turbine flow meter (TFM) and from the present gamma-ray meter. Total mass flow rate from both the upstream meters are plotted against the measurements from the newly developed venturi system in Fig. 14.

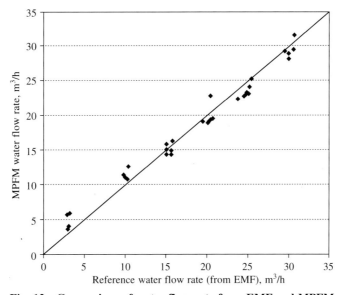

Fig. 12 Comparison of water flow rate from EMF and MPFM.

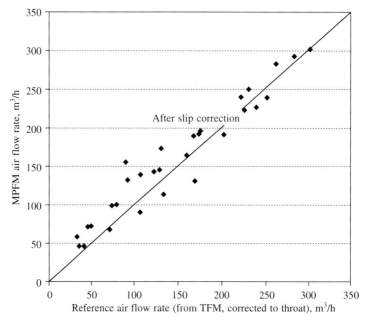

Fig. 13 Comparison of air flow rate from TFM and MPFM.

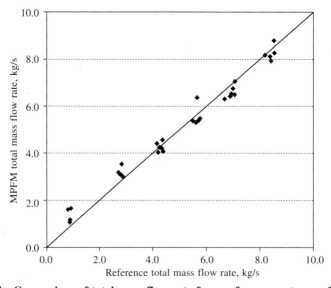

Fig. 14 Comparison of total mass flow rate from reference meters and MPFM.

5. Conclusion

The multi-phase flow meter predicts the total mass flow rate to an accuracy of better than ± 5.0%, volume flow rates of liquid and gas at accuracies of ± 6.0% and ± 8.0%, respectively, for the gas phase-fraction ranging from 60 to 99%. The comparison of the area averaged phase fraction (reconstructed by using Abels inversion method) with the measured data shows a deviation of 3%. The developed multi-phase flow meter system is intended for use in multi-phase flow conditions, wherein the phase-fraction variations are not so instantaneous, with a preset time of 5.0 sec with the specified accuracy levels.

References

1. Development of multi-phase flow meter using gamma-ray attenuation technique in conjunction with venturi, FCRI Internal Report FCRI/MPFM/01/2002.
2. H. Van Santen, Z.I. Kolar and A.M. Scheers, Photon energy selection for dual energy gamma and/or X-ray absorption composition measurements in oil-water-gas mixtures, *Nucl. Geo. Physics*, **9**, No. 3, 1995, pp.193–202.
3. Evaluation of static mixers for preconditioning of multi-phase flow, Christian Michelsen Research Report, CMR-94-A10017-Open.
4. F. Sanchez Silva, P. Andreusi and P. Di Marco, Total mass flow rate measurements in multi-phase flow by means of a venturi-meter, *Multi-phase Production*, 1991, pp. 144–155.
5. Glenn F. Knoll, Radiation Detection and Measurement, Third Edition, John Wiley & Sons Inc.
6. Engineering Outline: Jet Pumps, BHRA, Engineering Ltd, 3 May 1968.
7. Ahmed B. Al-Taweez and Steve G. Barlow, Field Testing of Multiphase Meters, *Saudi Aramco Journal of Technology*, Spring 2000.
8. Shollenberger, Gamma densitometry tomography of gas holdup spatial distribution in industrial-scale bubble columns, *Chemical Engineering Science*, **52**, No. 13, pp. 2038–2049.
9. Vest C.M., Tomography for properties of materials that bend rays: A tutorial, *Applied Optics*, **24**, No. 23, pp. 4089–4094.

Computerized Tomography for Scientists and Engineers
Edited by P. Munshi
Anamaya Publishers, New Delhi, India

5. Bone Imaging Using Compound Ultrasonic Tomography

P. Lasaygues[1] and P. Laugier[2]

[1]Laboratoire de Mécanique et d'Acoustique UPR CNRS 7051, 31 Chemin Joseph Aiguier,
13402 Marseille Cedex 20, France

[2]Laboratoire d'Imagerie Paramétrique UMR CNRS 7623 Université Pierre et Marie Curie,
15 rue de l'Ecole de Médecine, 75006 Paris, France

Abstract

The goal of ultrasonic tomography is to reconstruct the spatial distribution of some acoustic parameter of an object using ultrasonic measurements. The measurements are made for a set of emitter-receiver positions and of frequencies of the interrogating wave. The method is based on a linear approximation of the inverse scattering problem, the inverse born approximation (IBA), which allows to reconstruct small perturbations from a known-reference medium. For media with weak inhomogeneities, such as soft biological tissues (weak scatterers), the method works straightforward leading to a "constant background" IBA method, whose practical solution results in regular angular scanning with broad-band pulses, allowing to cover slice-by-slice the spatial frequency spectrum of the imaged object. This leads to "reconstruction-from-projections" algorithms like those used for X-ray computed tomography. For media with strong heterogeneities as bone (strong contrasts, large objects with respect to wavelength), the problem is non-linear and there is in general no single solution. In this case, ultrasound tomography suffers from an important limitation due to strong wavefronts distortion (refraction, attenuation) resulting in a poor quality of tomographic images. One solution is to compensate for refraction based on an exact/a priori/knowledge of the geometry and of the speed of sound of the object and on the assumption that the Snell's law is valid for each single ray. The exact shape and size of the test object can be reconstructed from initial measurements performed using ultrasonic reflection tomography. Ultrasonic transmission tomography, using compensation procedure, then provides quantitative values of the sound velocity. In case of inaccurate/*a priori*/object size and sound velocity, the algorithm becomes an iterative process which converges quickly towards the solution. Preliminary experimental results on human femur demonstrate the feasibility of quantitative images using compound ultrasonic tomography.

1. Introduction

Some bone pathologies like osteoporosis are directly related to bone remodeling dysfunctions leading to low bone mass, alterations in material properties or microarchitecture, and ultimately impaired bone strength. X-ray densitometry techniques represent the gold standard for assessing skeletal status and provide accurate and precise measurement of bone mineral density (BMD) [1]. Although BMD accounts for most of the variability of bone strength, several other bone properties also influence bone strength, such as material properties and micro architecture. Recent development of high resolution MRI and CT imaging techniques permits an estimate of several structural parameters of cancellous bone and cortical bone. Because ultrasound involves compressive and shear stress, quantitative ultrasound (QUS) has been introduced as an alternative to X-ray densitometry to probe multiple bone properties that contribute altogether to bone strength. QUS has been applied to the measurement of peripheral skeletal sites and many studies report the ability of QUS to prospectively predict fracture risk. However, interaction mechanisms of ultrasound with the structure of bone are not fully understood yet and the complexity of the phenomena has hindered the technical evolution of QUS and its widespread use in clinical settings. Several QUS measuring techniques are currently available, including transverse transmission to measure the heel bone (calcaneus) [2] and axial transmission which can be applied to multiple peripheral skeletal sites (radius, metacarpal, ulna, etc.) [3]. The basic principle of this technique relies on ultrasound transmission and estimates of the average value of the ultrasonic parameters (attenuation, sound velocity) for the wave pathway through the skeletal site. Other approaches exploit the reflection of ultrasound from cortical bone (ultrasound critical reflectometry) [4] or the signal backscattered from cancellous bone microarchitecture [5] in order to provide a direct assessment of material properties or micro architecture features. Ultrasound tomography represents an alternative to reach one or more significant parameters like acoustic impedance, ultrasonic velocity and attenuation. In a previous study, our group has shown that preliminary *in vitro* quantitative images of sound velocity in a human femur cross section could be reconstructed by combining ultrasonic reflection tomography (URT) and ultrasonic transmission tomography (UTT). The dramatic impedance mismatch between soft tissue and bone causes refraction, an effect which must be taken into account and compensated before the straight-ray reconstruction by the backprojections technique can be applied to UTT data acquisition. We have proposed to use URT first to reconstruct the external and internal boundary of cortical bone, and to modify the acquisition scheme for UTT according to the shape of the object in order to compensate for refraction.

2. Quantitative Ultrasonic Tomography (QUT): Distorted Born Iterative Method

Ultrasonic tomograph is brought about by a linearization of the inverse acoustic scattering problem. For soft biological tissues, the reference medium is water, *in vitro*, and the mean medium (the soft tissue) *in vivo*. In this case, the approximations of weak scattering are generally used, for example the approximation of Born or Rytov. On the other hand, if the medium is much more contrasted than the surrounding reference medium, the problem is non-linear and there is in general no single solution. This is the case for bones. Our idea is to adapt the methodology and inversion algorithms developed for soft tissues to bone (or solid tissue). This approach, analogous to the distorted born iterative method, relies on two assumptions, viz. hypothesis relative to the bone structure (and thus essentially to the choice of the site studied) and the hypothesis relative to the propagation of a wave in this kind of heterogeneous media.

We use the geometrical theory of diffraction, and more simply, a mode of propagation in straight beams.

2.1 Hypothesis and Assumptions: Distorted Born

The overall architecture of bone is divided into cancellous bone and cortical bone. Cortical bone forms a compact shell around highly porous cancellous bone, which is formed by interconnecting lattice work of bone filled with marrow. In general, cancellous bone is found in the axial skeleton (spine), small bones of the peripheral skeleton (e.g. calcaneus) and distal parts (epiphysis) of long bones such as radius and femur, while the diaphysis of long bones is composed primarily of cortical bone (e.g. radius, femur, tibia). This study is focused on cortical bone.

Considering the diaphysis (cortical) as a weakly heterogeneous, we can suitably apply, inside the medium, a linearization of the propagation and a "fluid" modeling. Cortical bone is generally assimilated to a transverse isotropic medium, with a principal axis of symmetry oriented along the longitudinal axis of the diaphysis. This assumption was confirmed by several independent measurements of compression and shear wave velocities in various cortical bone specimens and in various directions of isotropy [6, 7].

The correction of refraction on the water/bone and bone/water interfaces allows us to consider the scatterer as a weakly heterogeneous medium, consistently with Born's approximation.

Classically, there are two kinds of ultrasonic tomography, viz. (i) ultrasonic reflection tomography (URT), based on the measurement of the fluctuation of acoustic impedance, yields to the external or internal geometry of the object and (ii) ultrasonic transmission tomography (UTT), based on the measurement of the relative fluctuation of time-of-flight (TOF), provides the value of the sound velocity along the propagation in the object (Fig. 1). The relative fluctuation of velocity will be calculated by considering the TOF along a straight line of the acoustic wave in the cortical zone. TOF is directly related to the weak disturbance of the refraction index. Consequently, the velocity map may be obtained by inversing the TOF data.

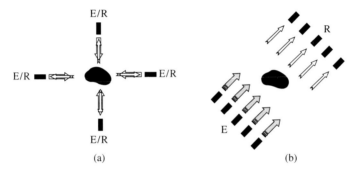

Fig. 1 Ultrasonic tomography: (a) reflection (URT) and (b) transmission (UTT).

2.2 Bone Shape Obtained from Ultrasonic Reflected Tomography Data

Ultrasonic reflected tomography gives a qualitative image of the shape of the body. At this level, the algorithm developed for the low-contrast media can be applied without restriction related to the material. Lasaygues and Lefebvre [4] showed that URT with signal processing [8] improves the resolution by eliminating interferences (speckle) and thus makes it possible to distinguish various boundaries from the objects. The external boundary of the object can easily be determined after URT image reconstruction. For the inner boundary, the problem is more complicated because we must determine the "apparent" internal border, which does not correspond to the actual one. Indeed, the reconstruction algorithm assumes the medium to be slightly diffracting with uniform sound speed (i.e. sound speed in water) as an initial reconstruction parameter. Consequently, the reconstructed object is deformed and the thickness of the cortical shell is incorrect.

Accurate positioning of the inner cavity boundary is reached by shifting the position of the "apparent" internal boundary previously measured. This correction of distance uses the *a priori* wave velocity in the medium (c_{bone}) derived from literature. The first step of the algorithm consists in representing the URT image in the cartesian plane mm/degree (Fig. 2). The second step determines the position of the external border (z_{ext}) and then of the internal border (z_{intl}). For modeling of the forward URT problem, the host-middle is water (c_{water}). As a consequence the new internal border location is determined relative to this reference velocity:

$$z_{int2} = \frac{c_{bone}}{c_{water}} [z_{intl} - z_{ext}]$$

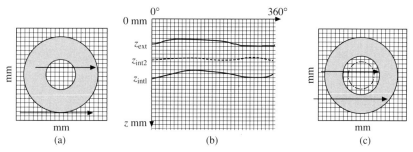

Fig. 2 Polar and cartesian representations of a URT: (a) initial image, (b) delimitation of the boundaries and (c) final image.

2.3 Correction of Refraction Phenomena for Ultrasonic Transmission Tomography

UTT is perturbed when the media are strongly contrasted. Basic assumptions on the tomography are also not violated and several phenomena appear. In order to apply to bone tomography the weak scattering tomography reconstruction algorithms, straight-line propagation must occur inside the material [9, 10], which requires the correction of waves refraction on water/bone and bone/water interface.

The compensation procedure may be achieved based on *a priori* knowledge of the geometrical shape and acoustical properties of the object [11]. Using Snell's laws we can compute the appropriate position and orientation of the transducers for straight parallel ray propagation in the object. Approximating the femoral shaft to a hollow tube, we use the refraction correction everywhere except in the hollow part (medullar canal), considered as a multi-layer medium with parallel and planar interfaces (Fig. 3). Then we define an original construction procedure to acquire acoustic sequences, so-called *two-zone correction procedure*.

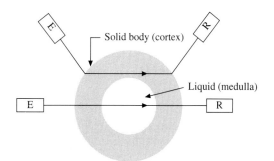

Fig. 3 Ultrasonic pathways for two-zone correction procedure for time-of-flight detection zone (zone 1: cortex, zone 2: cortex-medulla-cortex).

2.4 Iterative Quantitative Ultrasonic Tomography Algorithm

Quantitative ultrasonic tomography is the combination of the images obtained with URT and UTT. URT gives the size and shape of boundaries, and UTT assigns to each reconstructed pixel a sound speed value.

In case of inaccurate *a priori* information, the process described above may be an iterative one. Each QUT gives the geometrical and acoustical mean parameters of the next iteration. The complete iterative algorithm presented in Fig. 4 operates as follows:

URT is performed first and the shape of the object is reconstructed using some *a priori* sound velocity value (i.e. from literature or acoustic measurements). The external boundary is determined. For each view angle the apparent acoustic diameter is computed (i.e. *a priori* geometry). Using *a priori* velocity knowledge, the internal boundary of the object is detected (tube case). Second, using this reconstructed object as an *a priori* information for UTT, the UTT acquisitions are

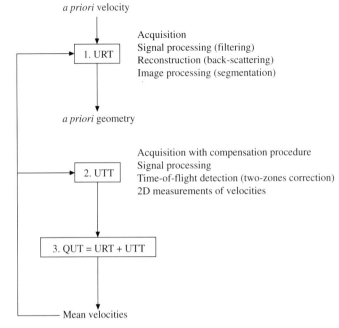

Fig. 4 General algorithm for quantitative ultrasonic tomography (QUT).

conducted following the algorithm of correction of refraction. TOF is determined from UTT data according to the two-zones. A TOF projection for each view angle is obtained and the corresponding transmission tomography image is finally reconstructed. Ultimately, we build the QUT.

3. Examples

3.1 Measurement System

The general architecture of the mechanical system is that of a first-generation X-ray tomograph: a main symmetric arm holds two transverse arms allowing the parallel translation of two transducers (1 MHz). Angular scanning is allowed by the rotation of either the main arm or the object holder. The transducers can also be positioned and oriented with high precision, allowing linear and sectorial scanning. Two 1 MHz ultrasonic transducers were connected to a computer via a digital oscilloscope. The axial resolution was 6 mm (4 ms) and the lateral resolution was 4 mm (Fig. 5).

3.2 Non-circular Test Object

Our first test object is a non-circular PVC tube (longitudinal wave velocity = 2700 m/s) (25 × 31 mm for external dimensions and 10 mm internal diameter). The object was placed in the center of the measurement system and 90 projections were acquired with 128 transverse displacements in steps of 330 μm (Fig. 6).

In this case, the complete iterative algorithm was used. On URT, we can extract the required (or pre-requisite) geometrical information to correct the refraction and to detect the TOF. The TOF extraction and the quantitative reconstructed image of ultrasonic velocity in the material are then computed.

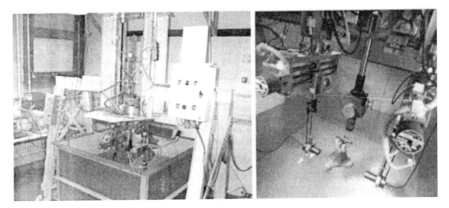

Fig. 5 Mechanical ultrasonic 2D-scanner.

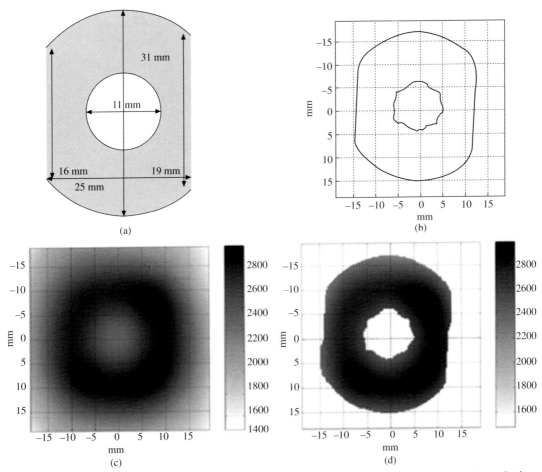

Fig. 6 Iterative QUT of a non-circular tube: (a) dimensions, (b) external and internal boundaries, (c) UTT with two-zone correction and (d) QUT = combined URT and UTT.

Good agreement was obtained between the reconstructed URT shape and the actual dimension of the object. The *a priori* velocity introduced was 2700 m/s and the mean velocity given on the QUT was 2742 m/s.

3.3 Human Femur

The bone sample was a human femoral specimen about 32 ± 5 mm in external diameter and 16 ± 2 mm in internal diameter.

For the *a priori* velocities, we chose 3400 m·s^{-1} in bone and 1478 m·s^{-1} in water. For the diameter, we initially compared the femur to a circular tube but the result was not satisfactory. By introducing a more realistic diameter, obtained for each view angle by URT, more accurate results were obtained (Fig. 7).

Fluid in the internal shape was reconstructed with a correct velocity value ($\simeq 1500$ m·s^{-1}) and the dimension of this cavity was 15-17 mm. The external diameter was found to be in the range 30-34 mm, close to the actual values (as shown on the X-ray computed tomography). Mean bone velocity of cortical shell was 3150 m·s^{-1}.

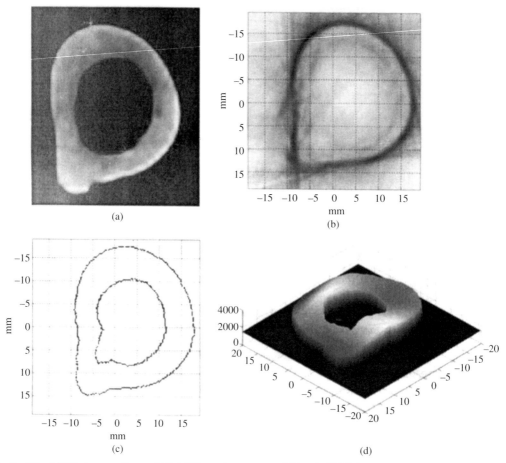

Fig. 7 The 1 MHz human femur UT: (a) X-ray tomogram, (b) basic URT, (c) acoustic boundaries plot and (d) QUT = combined URT and UTT with two-zone correction (sound velocity on vertical axis).

3.4 The 3D Representation

In Fig. 8 we compare two 3D US reflected tomograms of human female femur. The first one is obtained on a non-menopausal woman. The second is obtained on a menopausal osteoporotic woman. The cortical thickness is lower in the osteoporotic sample compared to the control one. This preliminary result suggests that geometrical characteristics, such as the cortical thickness, which is recognized as an important factor of fracture risk, can be derived from URT.

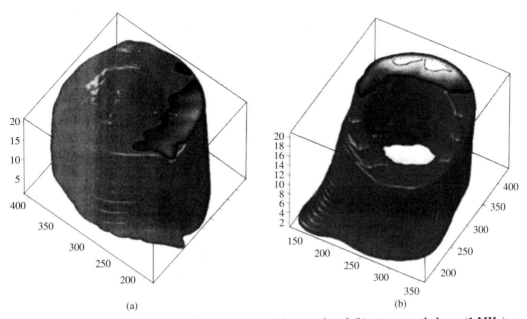

Fig. 8 Human femur (*in vitro*) without marrow: (a) normal and (b) osteoporotic bone (1 MHz).

4. Conclusion

An implementation scheme for bone ultrasonic imaging by ultrasonic tomography is presented. Because of strong contrast, the problem with bone is more complicated so we must change the usual framework of the tomography (linear inversion of the data with a linear approximation of the direct problem—Born approximation). It is necessary to study the non-linear inverse problem and an iterative method is inescapable (minimization of the distance between measurement and simulation of the forward problem). This approach depends on initial conditions (role of the *a priori* knowledge).

Our group studied three strategies in parallel. In this article we presented the progress of the first strategy based on Distorted Born Iterative method with iterative tomographic acquisition. The bone was approximated to a weak contrast object (i.e. weak local fluctuations of acoustic characteristics in a cross section of cortical bone) immersed in a quasi-homogeneous medium (water). An iterative transmission tomography was implemented with correction of wave refraction at bone/water interface. The preliminary results are encouraging.

References

1. Laugier, P., Challenges in quantitative ultrasound bone strength assessment: Status and perspectives. In: Computerized Tomography for Scientists and Engineers, edited by Prabhat Munshi, Anamaya Publishers, New Delhi, pp. 57-67 (2007).
2. Bossy, M. Talmant and P. Laugier, Three-dimensional simulations of ultrasonic axial transmission velocity measurement on cortical bone models, *J. Acoust Soc. Am.* **115** (5), pp. 2314–2324 (2004).
3. M.P. Andre, J.D. Craven, M.A. Greenfield and R. Stern, Measurement of the velocity of ultrasound in the human femur *in vivo, J. of Med. Phys.* **7**(4), pp. 324–330 (1980).
4. P. Lasaygues and J.P. Lefebvre, Cancellous and cortical bone imaging by reflected tomography, *Ultrasonic Imaging* **23**, pp. 55–68 (2001).
5. P. Lasaygues, E. Ouedraogo, J.P. Lefebvre, M. Gindre, M. Talmant and P. Laugier, Progress towards *in vitro* quantitative imaging of human femur using compound quantitative ultrasonic tomography, *Phys. Med. Biol.* **50**, pp. 2633–2649 (2005).
6. R.B. Ashman, Ultrasonic determination of the elastic properties of cortical bone: technique and limitations, PhD Thesis, Department of Biomedical Engineering, Tulane University (1982).
7. M. Pithioux, P. Lasaygues and P. Chabrand, An ultrasonic method to describe mechanical properties of compact bone, *Journal of Biomechanics* **35**, pp. 961–968 (2002).
8. Lasaygues, P., Lefebvre, J.P. and Mensah, S., Deconvolution and Wavelet Analysis on Ultrasonic Reflexin Tomography, *Topics on Non-Destructive Evaluation,* Series B. Djordjevic and H. Dos Reis, Series Editors, III International Workshop—Advances in Signal Processing for NDE of Materials, in X. Maldague, 3, Technical Ed, pp. 27–32, ASNT (1998).
9. McCartney, R.N., Jeffcott, L.B. and McCarthy, R.N., Transverse path of ultrasound waves in thick-walled cylinders, *Med. & Biol. Eng. & Compt.* **33,** pp. 551–557 (1995).
10. McCartney, R.N. and Jeffcott, L.B., Combined 2.25 MHz ultrasound velocity and bone mineral density measurements in the equine metacarpus and their *in vivo* applications, *Med. & Biol. Eng. & Comput.* **25,** pp. 620–626 (1987).
11. Ouedraogo, E., Lasaygues, P., Lefebvre, J.P., Talmant, M., Gindre M. and Laugier, P., Multistep compensation technique for ultrasound tomography of bone, *Acoustical Imaging* **26**, Kluwer Academic/Plenum, pp. 153–160 (2001).

Computerized Tomography for Scientists and Engineers
Edited by P. Munshi
Anamaya Publishers, New Delhi, India

6. Challenges in Quantitative Ultrasound Bone Strength Assessment: Status and Perspectives

P. Laugier

Laboratoire d'Imagerie Paramétrique UMR CNRS 7623 Université Pierre et Marie Curie,
15 rue de l'Ecole de Médecine, Paris, 75006 France

Abstract

Since the ultrasonic waves involve compressional or shear stress, the propagation characteristics of ultrasonic wave through bone are closely related to its mechanical properties (elastic constant) as well as to any other bone characteristics, such as bone mineral density, micro-architecture or micro-damages relating to mechanical properties. Quantitative ultrasound (QUS) measurements is playing a growing role in the assessment of skeletal status. Currently available technologies are based on measurements in transmission of the slope of the frequency-dependent attenuation and the speed of sound at peripheral skeletal sites (finger phalanges, heel, wrist). In this article, we will review the current state of development and outline a few promising developments.

Several investigations are currently being conducted to develop innovative QUS techniques, such as transmission imaging technology, reflection techniques based on backscatter, propagation of guided waves along cortical bone and ultrasound tomography. The feasibility of these different approaches and/or their clinical evaluation is currently being studied in different groups.

In contrast to X-ray based computed tomography which is a well-established technique in osteoporosis diagnostics, quantitative ultrasound tomography is still in its infancy and raises new challenges. In particular, an accurate interpretation of ultrasound measurement requires, first, a detailed understanding of ultrasound propagation with clear identification of the different waves and their exact propagation path that contribute to analyze signals. In other words, this requires solving the direct problem. Analytic solutions become rapidly inextricable due to the complexity of the medium. The situation is progressively being changed, with recent development of numerical tools based on finite difference modeling that permits to accurately model the propagation of ultrasound wave in 3D anisotropic models of bone. Numerical simulations can be performed on actual bone geometry, as measured from 3D X-rays computerized tomography (CT) for instance, combined with local effective elastic constants. Such computations have recently been able to elucidate the propagation of ultrasound in cortical bone. Another powerful advantage of these simulation tools is their potential to explore virtually different experimental configurations to measure bone properties (i.e. tomography) and therefore to guide engineers to design novel techniques for *in vivo* bone strength assessment.

1. Introduction

Today, reference methods for skeletal status assessment are based on X-ray absorptiometry techniques measuring bone mineral density (BMD). However, BMD is not sufficient to explain bone strength. Several bone properties such as microarchitecture or tissue elasticity which are not captured by conventional X-ray-based densitometric techniques also contribute to bone strength independently of bone mass. The alternative to X-ray is represented by quantitative ultrasonic (QUS) methods which have been introduced in 1984 [1], based on the cheapness of the technology and on the potential of elastic waves to probe bone material properties and microarchitecture. Like X-ray densitometry techniques, QUS has been adapted to assess different skeletal sites consisting of different types of bone (Figs. 1 and 2). Devices are currently available to measure easily accessible peripheral sites such as the heel, fingers, wrist or tibia (Table 1).

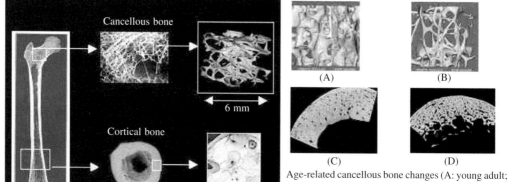

Age-related cancellous bone changes (A: young adult; B: elderly). Age-related cortical bone changes (C: young adult; D: elderly). Cortical bone becomes thinner and more porous. Cancellous bone becomes more porous, trabeculae are thinner and disconnected. Images provided by synchrotron radiation microtomography. Courtesy: Dr. F. Peyrin, ESRF, Grenoble

Fig. 1 **Structure of cancellous bone and cortical.**

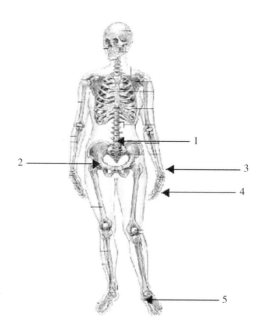

Fig. 2 Main fracture skeletal sites (1: femur; 2: spine; 3: wrist). Main skeletal sites for X-ray absorptiometric measurements (1, 2 and 3) and quantitative ultrasound (4 : finger phalanges; 5: heel).

Table 1. Technique diversity

Site	Bone Type	WB/nWB	Technique	Coupling	Image	Parameters BUA	Velocity
Calcaneus	TB	WB	TT	Wet	No	Yes	Yes
Calcaneus	TB	WB	TT	Wet	Image	Yes	Yes
Calcaneus	TB	WB	TT	Wet	Image	Yes	Yes
Calcaneus	TB	WB	TT	Dry	No	Yes	Yes
Calcaneus	TB	WB	TT	Dry	Adaptive	Yes	Yes
Phalanges	IB	nWB	TT	Dry	No		Yes
Tibia	CB	WB	AT	Dry	No		Yes
Multi-site	TB, CB	WB, nWB	AT	Dry	No		Yes

TB: trabecular bone; CB: cortical bone; IB: integral bone; WB: weight bearing bone; nWB: non-weight bearing bone; TT: transverse transmission; AT: axial transmission.

2. Technological Aspects

2.1 Transverse Transmission

The transverse transmission technique uses one ultrasound transmitter and the other acting as a receiver. Both transducers are placed on each side of the skeletal site to be tested along its mediolateral axis (i.e. widthwise). The method has been applied to the measurement of the slope of frequency-dependent attenuation (or BUA, dB/MHz) and the speed of sound (SOS, m/s) at calcaneus (500 kHz) and the amplitude dependent Ad-SOS at the finger phalanxes (1.25 MHz). The BUA represents the rate of loss of the signal which is transmitted through the bone as a function of frequency (Fig. 3).

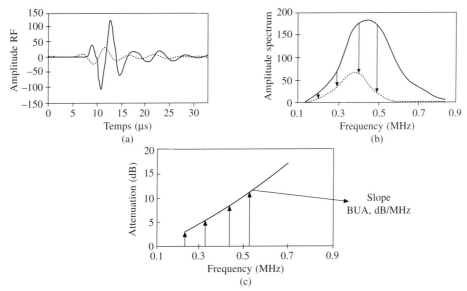

Fig. 3 (a) Typical examples of radiofrequency signals (reference: solid line; transmitted through the heel: dashed line, (b) amplitude spectra and (c) attenuation curve (spectra log ratio) plotted as a function of frequency.

The calcaneus is the most popular measurement site and the majority of clinical reports have focused on this bone. Calcaneal devices show the greatest diversity of technical approaches with fixed single point systems, moving transducers systems and imaging devices employing either water bath or ultrasonic gel coupling between the transducers and the skin [2]. QUS imaging systems provide the end-users with quantitative ultrasound images in transmission mode (Fig. 4) [3]. Spatial mapping may be employed to define one or more regions of interest with respect to anatomical landmarks of the bone of the subject or features of the topology of the diagnostic measure itself [4]. An average value of QUS parameters can be calculated by averaging the values at each measurement point in the region of interest (ROI) selected by the operator on the image or the ROI automatically placed by the software (Fig. 4).

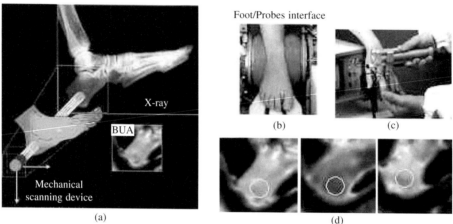

Fig. 4 (a) Heel BUA imaging using a mechanical scanning device, (b) heel BUA imaging using a pair of 2D transducer array in contact with the skin (Courtesy: Dr. M. Defontaine, LUSSI Tours), (c) QUS measurement at finger phalanges (Igea, Capri, Italy), (d) automatic circular regions of measurement in three different subjects.

With most of the commercial devices, the thickness of the heel is unknown. BUA and SOS will consequently be affected by bone size as well as bone material properties. It has been argued that inter-individual variability in heel width is small in practice and can be safely neglected [5] and that no significant improvement in fracture discrimination could be seen after normalization for bone width [6]. Bone width could become a significant cofounder in growing children or if different ethnic groups, males or females were to be compared. Typical range of SOS values in the heel range between 1450 and 1650 $m \cdot s^{-1}$.

There is one commercially available device which measures the amplitude dependent speed of sound (Ad-SOS) at the distal metaphysis of the first phalanx of fingers II-V [2]. In the metaphysis, both cortical and trabecular bone are present. The instrument is equipped with two transducers mounted on an electronic caliper that measures the distance between the probes (Fig. 4). The probes positioning at the distal metaphysis of the phalanx is slightly adjusted until the optimum signals are recorded and Ad-SOS can be measured. Measurements are carried out on each of the four phalanges and the results averaged. Typical range of SOS values through finger phalanges range between 1850 and 2250 $m \cdot s^{-1}$.

2.2 Axial Transmission

Ultrasonic measurement along a length of long bone has attracted the attention of a number of researchers. The axial transmission technique currently uses a set of transducers (transmitters and receivers) to measure ultrasound velocity along a fixed distance of the cortical layer of the bone, parallel to its long axis. This set of transducers is placed on the skin along the bone and measures the arrival time of the wave, which propagates along and just below the surface of the bone after entering it at the critical angle (Fig. 5). In contrast to transverse transmission techniques, which require to place a transducer on each side of the bone, the axial transmission technique, with its easy transducer set-up, may be applied to a greater number of skeletal sites. The time-of-flight of the first arriving signal is measured and used to calculate velocity. In commercial devices, the velocity may be calculated from either the transmitter-receiver distance divided by the time-of-flight or by dividing the known distance between two receivers by the corresponding difference in time-of-flight of the first signal.

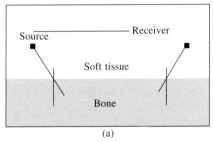 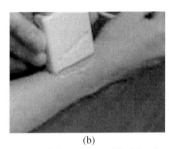

Fig. 5 Principle of axial transmission: (a) travel path of the wave guided by the cortical bone surface and (b) measurement on the forearm.

The theoretical analysis of the field reflected from a fluid-solid interface for an incident spherical wave predicts the existence of a lateral wave propagating along the sample surface at a velocity close to the longitudinal velocity, in addition to the ordinary reflected wave and vibration modes. For thin plate or thin cortical shell, dependence of velocity on thickness has been reported, suggesting that lateral longitudinal wave is not consistently the first arriving signal and that other propagating modes, such as plate modes may arrive first. One device is currently commercially available for the measurement (1.25 MHz) of several skeletal sites, including the radius, finger phalanges, tibia and hand metacarpal. Typical range of SOS values in cortical bone (measured along the long axis of bone) measured *in vivo* at various skeletal sites is 3500-4200 m·s^{-1} (radius, tibia, phalanxes, metacarpal).

2.3 Advanced Signal Processing

Such parameters like attenuation or speed of sound resulting from a simple but robust analysis is a reduction of the wealth of information hidden in transmitted signals. Better description of interaction mechanisms together with sophisticated signal analysis or new acquisition procedures could potentially lead to enhanced ultrasound assessment of bone strength. Such enhanced signal analysis has been tested for the phalanx and several parameters derived from the ultrasonic trace were derived such as fast wave amplitude, number of peaks, signal dynamic, bone transmission time, growing trend of the peaks amplitude, etc. Recent studies have pointed the values of some of these parameters to reveal information on structural features of the bone. Barkmann [8] in a study performed in human phalanges have found that cross-sectional cortical area, medullary canal area and relative cortical area could be calculated from the speed of sound and wave amplitude. A few *in vivo* data using this analysis carried out in postmenopausal women, men and on patients with hyperparathyroidism are reported in literature [9-11]. For example, the study by Wuster [9] showed that not only Ad-SOS but also parameters characterizing the transmitted signal at the phalanges could discriminate between osteoporotic and healthy women and predict fracture risk. In a pilot study on a limited number of subjects, Montagnani [10, 11] have reported that parameters such as fast wave amplitude and bone transmission time, but not Ad-SOS, could discriminate between different bone diseases such as osteoporosis and hyperparathyroidism. A new parameter, the *ultrasound bone profile index* (UBPI) results from such enhanced signal analysis. It is based on a combination of selected features of the ultrasonic signal specifically related to bone structural properties and processed in a statistical approach to express the probability level of incurring a fracture and to provide automatic computer assisted analysis [9].

2.4 Innovative Acquisition Procedures

Innovative acquisition procedures may also be helpful to grasp bone properties that cannot be easily assessed in transmission or to assess other skeletal sites. Currently available imaging devices measure the heel. Chappard [12] demonstrated the feasibility of QUS imaging at a different site (Fig. 6) with reasonable precision. The feasibility of imaging acoustic properties of the upper femoral extremity is currently under study in our group (Fig. 7). Reflection techniques such as ultrasonic backscatter have been introduced recently for the measurement of specific material and structural bone properties. In addition, skeletal sites, which are difficult to reach by transmission could be evaluated by such reflection techniques.

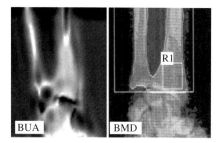

Fig. 6 BUA and BMD images of the wrist *(in vivo)*.

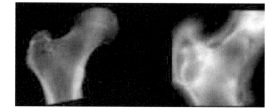

Fig. 7 BMD (left) and BUA (right) images of the femur *(in vitro)*.

Backscattering from bone has received much less attention than attenuation and speed of sound. Backscatter represents the re-radiation in the backward direction of an ultrasonic incident wave from small obstacles, e.g. trabeculae. The study of backscatter is important, however, because it can provide information regarding size, shape, number density, and material properties of obstacles. Backscatter measurements have been proposed for their potential to probe specifically bone microarchitecture. The clinical feasibility and diagnostic promise of this measurement have already been demonstrated [13-16]. The frequency-averaged backscatter coefficient between 0.2 and 0.6 MHz has been termed broadband ultrasonic backscatter (BUB). The relationship between BUB and fracture risk has been documented retrospectively [16] and found to be significant. Despite demonstration of their diagnostic promise, these new techniques with pilot studies or retrospective study validation only must await further prospective validations before they can be generalized to routine clinical utilization.

3. Practical Use

There is no standardization among manufacturers regarding either the experimental and signal processing procedures used to estimate the parameters or the way the results are presented to the physician. Sometimes, BUA and SOS values are expressed as Z- and T-scores to help the physician interpret the results. An additional output variable provided by several devices is a linear combination of BUA and SOS. Such a combination, defined empirically, compensates for temperature variation and offers better stability [17]. It is thought to have better precision and greater diagnostic capabilities than either parameter by itself. The actual coefficients of the linear combination between BUA and SOS resulting in a stiffness index or equivalent quantities (stiffness index or SI, quantitative ultrasound index or QUI, strength index or STI, ...) depend on the manufacturer. While ultrasound does not measure BMD directly, some clinical studies have shown that the QUS parameters at the heel are highly correlated with heel BMD as measured by DXA. Therefore, one device also yields an output variable referred to as estimated heel BMD (Est. BMD, g/cm^2).

Commercial QUS devices show great technological diversity (Table 1). Because there is no standardization among manufacturers of ultrasound densitometers, substantial differences in BUA and SOS values may be found between different commercial devices, depending on several device characteristics (e.g. transducers, ROI), measuring conditions (e.g. temperature control, image or single point measurement, dry or wet coupling) and signal processing algorithms. Until there is an accepted standard for ultrasound measurement in terms of signal processing, parameters definition and ROI, the results obtained on one device cannot be directly compared to those from another, even when both claim to deliver the same measurement (e.g. BUA, SOS, etc.) at the same skeletal site (e.g. calcaneus).

Despite the continuous efforts being made by most constructors to implement automatic measurement procedures and stringent quality control on their devices, it must be kept in mind that to some extent QUS measurements still remain operator-dependent and that the need for adequate specific training for each QUS device is mandatory before reproducible measurements can be obtained. In particular, the various causes of error, including coupling, soft tissue, temperature, positioning and movements must be known from the operators.

4. Clinical Aspects

Recent findings and reviews confirm QUS diagnostic sensitivity and clinical utility for the prospective assessment of fracture risk [2, 18]. Normal bone demonstrates higher attenuation, and is associated with greater velocity compared with osteoporotic bone. As for X-ray absorptiometry, the influence of age was observed with a more pronounced decrease immediately after menopause. Various retrospective studies have reported discriminating power of QUS between osteoporotic and age matched control similar to other X-ray densitometry techniques [2, 18].

The strongest evidence regarding the association between QUS and fracture risk comes from two large scale prospective studies of older women [19, 20]. These studies reported similar predictive values of QUS at the calcaneus and BMD measured at the femur for femoral or non vertebral fractures in elderly subjects of more than 65 years. The association between QUS parameters and fracture remained significant after adjustment for hip BMD. This result indicates that at similar hip BMD values, QUS measurements can further discriminate between individuals. To interpret this result, it has often been speculated that QUS and BMD measurements reflect different bone properties and consequently provide independent information regarding fracture risk. However, this interpretation is not strongly supported by recent experimental studies documenting the relationship of QUS measurements to bone mass, microarchitecture and strength [21, 22]. It is more likely that measurements at different skeletal sites partly reflect independent information owing to skeleton heterogeneity.

The value of ultrasound densitometry in younger women (i.e. perimenopausal women) is less certain, although the previous result was recently confirmed for a population of younger women (45-75 years) [23]. The association between ultrasonic measurements at the heel and vertebral fracture [24] and QUS measurements at the phalanges and non-vertebral fractures [25] have also been reported.

Finally, there is a strong correlation between BUA and bone density measurements at the same site ($r^2 = 0.8$) [26]. This suggests that BUA may be closely linked to bone density at the site of measurement. There have been limited investigations regarding the ability of ultrasound densitometry to measure changes over time. The value of serial measurements, particularly at the calcaneus, to follow response to therapy has been highly controversial. Inconsistent results have been reported between many therapeutic assays. There may be practical reasons to such an inconsistency: limited precision of the devices used in clinical trials, insufficient knowledge of factors influencing long-term reproducibility (external and physiological factors) resulting in poorly qualified measurement protocols. Furthermore, it has been argued that measurements from peripheral skeletal sites will not be useful for monitoring since, they do not reflect treatment-related bone changes that are observed at central sites, particularly the spine. A number of studies report on QUS use in other skeletal diseases such as endocrine, renal, or genetic disorders, drug-induced osteoporosis (glucocorticoid), genetic diseases, immobilization, pregnancy and lactation, loco-regional osteoporosis associated with arthritis, chronic and inherited disorders in childhood [2].

Several clinical issues that need to be addressed are as follows:

(a) Although many retrospective and prospective studies have consistently reported the predictive value of QUS measurements for fracture risk, there is still no consensus on how to use QUS measurements to manage individual patients; age-related decrease of the T scores depend on the technique and the skeletal site and serious concerns have been expressed on using the WHO criteria of osteoporosis (T score < -2.5) for QUS techniques.

(b) Does combining measurements at several sites enhance fracture risk prediction? Preliminary clinical validation of axial transmission technique applied to several sites (radius, phalanx, metacarpal, calcaneus) has been reported. Discrimination of osteoporotic fractures at all sites was statistically significant and moderate improvement in the discriminative ability was showed from the combination of several sites [27, 28].

(c) Does combining several parameters enhance fracture risk prediction? Only one study has reported that the combination of backscatter and transmission measurements at the calcaneus did not perform better than either parameter alone [16].

5. Theoretical and Experimental Aspects

The nature of the information on bone properties conveyed by ultrasonic measurements has been a central question since many years. The research field has produced a significant number of experimental studies looking at the relationships of a number of bone properties (e.g. BMD, microarchitectural features or biomechanical properties) with QUS parameters. Strong correlations between ultrasonic parameters and mechanical properties (Young's modulus or strength) are reported for femoral and calcaneal specimens. The best prediction of the mechanical properties is obtained with multiple regression models combining SOS and BMD [29, 30]. In contrast,

BUA does not provide independent prediction compared to BMD alone [31]. Cadavers studies compared ultrasonic measurements at the calcaneus with the bone strength of matched calcaneus [32], femur [33] or vertebrae [34]. The correlations are strong for the calcaneus, moderate for the femur and weak but significant in the case of the vertebrae. Regarding bone strength, the predictive value of QUS at the calcaneus is comparable [32, 34] or weaker [35] than the prediction obtained with BMD alone, and again, the combination of the various quantities (BMD and BUA or SOS) does not improve the prediction of strength for the various sites. These results suggest that current QUS measurements may be considered as surrogate for BMD at the tested skeletal site.

It is virtually impossible to model by analytic means the extremely complex field resulting from the interaction of an incident wave with cortical or cancellous bone taking into account the full complexity of the geometry, boundary conditions, anisotropy and inhomogeneity in bone properties. Recently developed simulation methods based on finite difference offer a fertile alternative to inextricable analytic formulations. Such wave propagation simulation has been applied to the problem of axial transmission along the radius [35] and transverse transmission through the phalanx [9]. In each case, modelization has been found to be of great value in giving insight into the properties (nature, pathway) of propagating waves. It can be used to elucidate the relationship between SOS and bone properties (elasticity, porosity, cortical thickness, cortical area, etc.) and test inverse calculation procedures. Axial transmission SOS is deterministically determined by the material properties of the cortex (mineralization, porosity) as well as by the cortical thickness [36, 37]. Finite difference numerical simulations of ultrasound propagation through a model of the cross-section of a human phalanx were very helpful to document the influence of several bone structural parameters on quantitative measurements of Ad-SOS [9]. An example of snapshots illustrating the transmission of a plane wave through a cylindrical model of the phalanx is shown in Fig. 8. In summary, while the velocity of the fastest signal mostly depends on the cortical cross-sectional area, the amplitude depends mostly on the area of the medullary canal. These results open interesting perspectives for the development of clinically useful data inversions schemes in which material and morphological properties are derived from measurements of transmitted signals. Another powerful advantage of these simulation tools is their potential to explore virtually different experimental configurations to measure bone properties (i.e. guided wave, nonlinear acoustics, ultrasound tomography) and therefore to guide engineers to design novel techniques for *in vivo* bone strength assessment. The acoustic modelization of cancellous bone is much more complex owing to the greater complexity of this material.

6. Conclusion

There is considerable research interest in QUS bone measurements. Several investigations to develop innovative QUS techniques to determine and utilize the full potential of QUS are currently conducted by several groups. There is substantial diversity in QUS approaches (diversity in technological approaches, algorithms, skeletal sites), which limits comparability among devices and indicates that QUS has not reached maturity yet. However, the potential of widespread use of this technology to assess skeletal status in primitive osteoporosis and many other bone-related diseases is huge. QUS measurements are predictive of future fracture risk. However, several clinical issues still remain unclear such as management of individual patient or monitoring capabilities. Because ultrasonic waves involve compressional or shear stress, the propagation characteristics of ultrasonic wave through bone are closely related to its mechanical properties (elastic constant) as well as to any other bone characteristics, such as bone mineral density and

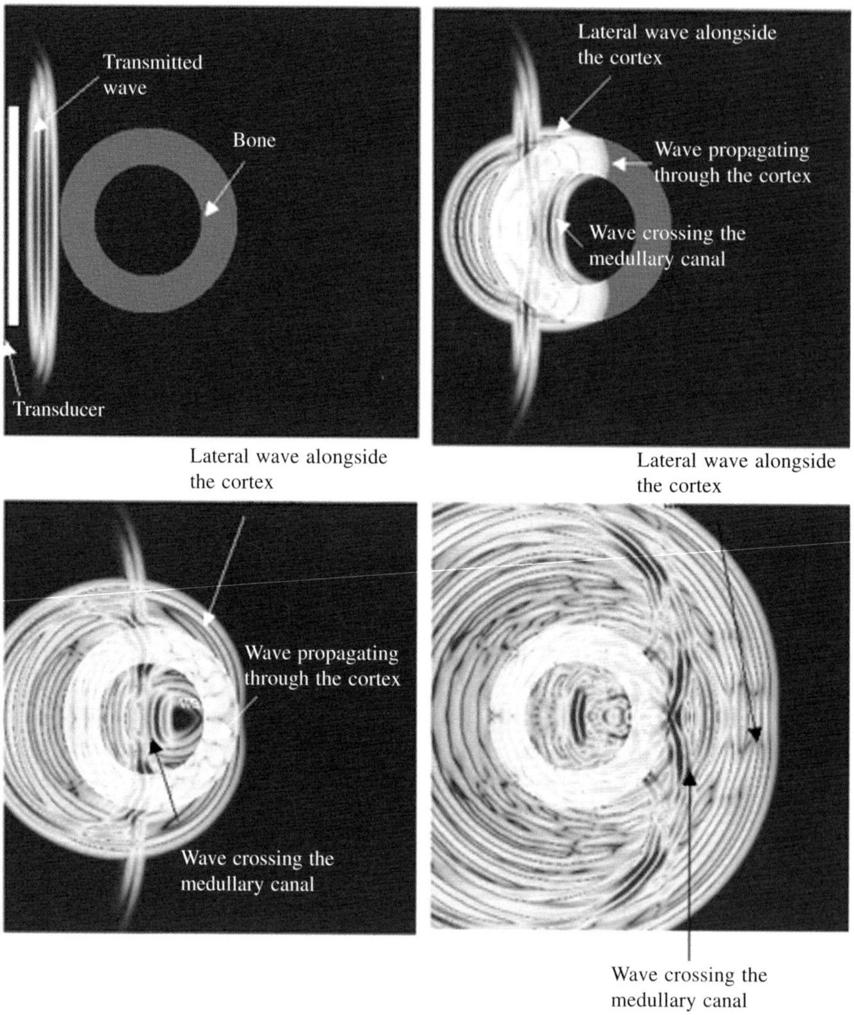

Fig. 8 Transmission of a plane wave through a cylindrical model of the phalanx finite difference modeling clearly showing several pathways through the phalanx.

microarchitecture, relating to mechanical properties. For cortical bone, recent theoretical and experimental approaches have improved our understanding of the underlying interaction mechanisms between ultrasound and bone. SOS assessed in axial transmission at the radius reflects changes in tissue mineralization and cortical porosity. For cancellous bone, theoretical approaches must be refined to provide better knowledge about what is measured.

References

1. Langton, C.M. *Eng Med.* **13**: 89–91 (1984).
2. Quantitative ultrasound: Assessment of osteoporosis and bone status. Eds. London: Martin Dunitz (1999).

3. Gregg, E.W. *Osteoporosis Int.* **7**: 89–99 (1997).
4. Laugier, P. *Calcif Tissue Int.* **58**: 326–331 (1996).
5. Fournier, B. *Osteoporosis Int.* **7**: 363–369 (1997).
6. Wu, C.Y. *Bone* **16**: 137–141 (1995).
7. Blake, G.R. *Br. J. Radiol.* **67**: 1206–1209 (1994).
8. Barkmann, R. *Osteoporosis Int.* **11**: 745–755 (2000).
9. Wuster, C. *J Bone Miner Res.* **15**: 1603–1614 (2000).
10. Montagnani, A.S. *Osteoporosis Int.* **11**: 499–504 (2000).
11. Montagnani, A.S. *Osteoporosis Int.* **13**: 222–227 (2002).
12. Chappard, C. *Osteoporosis Int.* **11**: S33 (2000).
13. Laugier, P. IEEE Ultrasonics Symp. 1101–1103 (1997).
14. Roberjot, V. IEEE Ultrasonics Symp. 1123–1126 (1996).
15. Wear, K.A. *Ultrasound Med Biol.* **24**: 689–695 (1998).
16. Roux, C.*J Bone Miner Res.* **16**: 1353–1362 (2001).
17. Chappard, C. *Osteoporosis Int.* **9**: 318–326 (1999).
18. Gregg, E.W. *Osteoporosis Int.* **7**: 89–99 (1997).
19. Hans, D. *Lancet.* **348**: 511–514 (1996).
20. Bauer, C. *J Bone Miner Res.* **10**: S175 (1995).
21. Nicholson, P.H.F. *Bone* **23**: 425–431 (1998).
22. Chaffaî, S. *Bone* **30**: 229–237 (2002).
23. Thomson, P.W. *J Clin Densitometry* **1**: 219–225 (1998).
24. Huang, C. *Calcif Tissue Int.* **63**: 380–384 (1998).
25. Mele, R. *Osteoporosis Int.* **7**: 550–557 (1997).
26. Chappard, C. *Osteoporosis Int.* **7**: 316–322 (1997).
27. Hans, D. *J Bone Miner Res.* **14**: 644–651 (1999).
28. Barkmann, R. *J Clin Densitometry* **3**: 1–7 (2000).
29. Njeh, C.F. *Osteoporosis Int.* **7**: 471–477 (1997).
30. Hodgkinson, R. *Bone* **21**:183-190 ; (1997).
31. Langton, C.M. *Bone* **18**: 495–503 (1996).
32. Han, S. *Calcif Tissue Int.* **60**: 21–25 (1997).
33. Bouxsein, M.L. *Calcif Tissue Int.* **56**: 99–103 (1995).
34. Cheng, X.G. *J Bone Miner Res.* **12**: 1721–1728 (1997).
35. Bossy, E. *J Acoust Soc Am.* **112**: 297–307 (2002).
36. Bossy, E., Peyrin, F., Talmant, E., Laugier, P. *J Bone Miner Res*, in press (2004).
37. Bossy, E., Talmant, M., Laugier, P. *J Acoust Soc Am*, in press (2000).

Computerized Tomography for Scientists and Engineers
Edited by P. Munshi
Anamaya Publishers, New Delhi, India

7. Computerized Tomography in Blast Furnace

S.K. Mandal
Engineering Division, National Metallurgical Laboratory, Jamshedpur-831 007, India

---

Abstract

Many experimental studies have been carried out to study the workings of packed bed reactors such as blast furnace, catalytic reformer and solid drying process using various two/three dimensional models. Blast furnace is the most complex and typical. However, no experimental data are available in the literature on the estimation of permeability distribution in a packed bed, which is very essential and important to understand the dynamics of the packed bed reactors.

The main idea is to create prototype facilities for the estimation of permeability distribution in a packed bed situation similar to blast furnace at cold condition through computerized tomography using X-ray radiographic system. This will help in three-dimensional visualization of the interior of a packed bed situation through the image reconstruction from the image slices obtained from X-ray radiography system.

The systems involving X-ray, gamma ray, ultrasonic, eddy current etc. are principally used in non-destructive testing (NDT). Most of the methods used in these systems can be combined into the mathematical techniques of computerized tomography (CT). The mathematical techniques used are based on convolution back projection algorithm (CBP).

---

1. Introduction

The conventional film technique in the field of radiography is fast changing with the advent of high-speed digital technology in the fields of radiography and computerized tomography [1, 2]. Various analytical techniques are available in order to achieve quantitative evaluation of interior inspection and quality control of materials and components. Mathematical modelling and simulation may play an important role in the qualitative as well as quantitative evaluation of radiographic images.

Packed beds are generally used in chemical and metallurgical applications such as blast furnace, catalytic reformer, solid drying process [3] etc. The sub-optimal performance of the blast furnace (Fig. 1) in India is due to the following reasons:

(1) Majority of the blast furnaces are very old
(2) Performance tuning is done by heuristic methods only
(3) Instrumentation is inadequate
(4) There is automatic control
(5) R&D is scattered and non-integrated
(6) There is no national level linkage of R&D

In view of the above, many theoretical models are developed to verify the kinetics of blast furnace [4, 5] and various R&D efforts/attempts are still on to improve its productivity. National/international scenario is given in Table 1. Burden distribution model, flow model, thermo-chemical model, raceway model, freeze line/open hearth model are the most important building

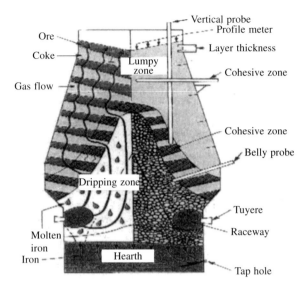

Fig. 1 Iron making blast furnace.

Table 1. National/international scenario

Parameter	National	International
Productivity (tonne/metre^3/day)	0.9-1.9	2.2-2.8
Fuel rate (kg/tonne hot metal)	575-750	450-500
Coal injection (kg/tonne hot metal)	70-125	150-200
Silicon (percentage)	0.5-1.5	0.1-0.3
Sinter (percentage)	50-70	80-85
Blast temperature (degree centigrade)	1000-1250	1100-1350

Each of 30 blast furnace's production is 20 tonnes per annum.

block models for dynamic study of blast furnace. Permeability distribution aspects in burden distribution model play an important role in the kinetic study of blast furnace. In view of this, simulated type of facilities are being created for the qualitative and quantitative estimation of permeability distribution in a packed bed situation similar to blast furnace at cold condition through computerized tomography using X-ray radiographic system [6, 7]. This will help in three-dimensional visualization of the interior of a packed bed situation through the image reconstruction from the image slices obtained from X-ray radiography system.

This article deals with the creation of a typical X-ray radiography facility at NML which will be the only facility available in the Eastern India region and attempts to develop open ended facility for the estimation of permeability distribution in a packed bed situation similar to blast furnace reactors at cold condition in the first phase through the processing of radiographic image slices in each scan. In future, all those image slices may be used for reconstruction of three-dimensional image of the whole packed bed for quantitative evaluation of the porosity and permeability distribution in it.

2. Facility Configuration

Proposed X-ray radiography system facility is shown in Fig. 2. It comprises X-ray generator as source along with mechanical slit assembly, object manipulator and detector; all interfaced with computer. The system is used for overall cone beam X-ray tomography [8, 9]. Special features of each are briefly described in Table 2.

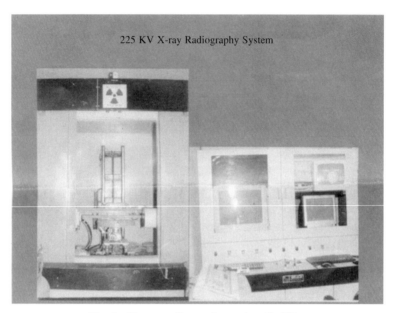

Fig. 2 X-ray radiography system facility.

The X-ray beam falls on the cylindrical sample holder acting as packed bed simulated similar to blast furnace rested on the turn table containing iron ore as sample under test for estimation of permeability distribution. The X-ray beam after attenuation through the sample under test falls on the detector. The visible radiation corresponding to X-ray intensity is converted into video signal with the help of mirror and CCD camera combination, which is interfaced with computer. The amount of energy of X-ray requirement and the acquisition time per scan is type of material and its thickness dependent. The entire geometry of the sample is scanned through rotation of the turn table on which sample is rested.

3. Principles of Reconstruction

X-ray radiography system acts as scanner and it is assumed that parallel beam of X-rays are infinitely thin. If $f(x, y)$ is the absorption coefficient of objects at a point x, y in a slice then the detected beam is related as

$$i = i_0 \exp[-\int_L f(x, y) \, du]$$

where L is the path of X-ray, u the distance along L, i_0 the intensity of the unattenuated X-ray beam and i the intensity of the attenuated X-ray beam after it has traversed a layer of material with definite thickness [10]. The linear attenuation coefficient exhibits dependency on mass attenuation coefficient and mass density. However, Lambert-Beer law of absorption applies only

Table 2. X-ray system

Technical data	ISOVOLT 225 HS
High voltage generator	
Primary frequency (kHz)	40
Maximum power anode/cathode (kW)	3/3
Maximum current (Ma)	45
High voltage ripple (V/mA)	12
Generator weight anode/cathode (kg)	123/170
Generator insulation	Oil
Power module	Three phase 415 V
Control module	
Number of storable programs	250
X-ray tube	
High voltage steps (kV)	0.1 from 5 to 225
Tube current steps (mA)	0.1 from 0.1 to 45 Ma
Repeat accuracy kV (%)	± 0.01
Repeat accuracy mA (%)	± 0.01
Prewarning (sec)	2-250
Movement	X, Z and tilt ±15°
Object manipulator	
Technical data	*Parameter*
Turn table base diameter	600 mm
Rotation	360°
Movement	Lateral
Support weight of object	Up to 200 kg
Speed of rotation detector	Adjustable (2° per min normal)
Technical data	*Parameter*
Real time image intensifier	Cesium iodide input detector
Screen size	9 inch
Distortion	< 1% at the centre and 2% at the edge
Energy range	10-225 kV
QDE	> 60%
Resolution	
Normal	5.2 lp/mm
Mag I	5.8 lp/mm
Mag II	6.4 lp/mm
CCD camera	High resolution camera
Movement	X, Z and tilt of ±15°
Shutter	5 mm thick lead motorized

to monoenergetic X-rays, a perfectly homogeneous material thickness, but typical industrial applications normally use polychromatic radiation and, additionally, the materials under investigation are often far from homogeneous.

Now observed signal is defined as:

$$g = \ln(i_0/i)$$

Hence, the linear transform will be as

$$g(s, \theta) = \int_L f(x, y)\, du,\ -\infty < s < \infty\,;\ 0 \leq \theta < 2\pi$$

where (s, θ) is the coordinate of the X-ray relative to the object.

Now radon transform is defined as

$$g(s, \theta) = r f(x, y)$$

where $$f(x, y) \to g(s, \theta) \to \bar{g}(s, \theta) \to f(x, y).$$

4. Design of Experiments and Calibration

At the beginning, the design of experimentations are sequentially done as follows:

(a) Test the system with homogeneous plastic balls
(b) Test the system with non-homogeneous plastic balls
(c) Test the system with homogeneous metal sample
(d) Test the system with non-homogeneous metal sample
(e) Test the system with heterogeneous metal samples
(f) Calibrate the situation with known sample [11]

5. Results and Discussion

The X-ray images in Figs. 3(a), 4(a) and 5(a) are the image projections at angular positions of 0°, 90° and 180°, respectively. These images are the images of cylindrical sample holder containing iron ore. The size of the cylindrical sample holder is 50 mm diameter and 150 mm height. The average particle size of iron ore is 10 mm diameter. The images in Figs. 3(b), 4(b) and 5(b) are the detected images of Figs. 3(a), 4(a) and 5(a). This is done using split and merge algorithm [12, 13]. These images clearly show the distinct porosity distribution in the packed bed under test along with interconnections qualitatively [14, 15, 16]. The porosity distribution could have been shown clearly if 3D image is reconstructed using various number of projections [17, 18, 19]. Main causes of imperfect experimental data in general are due to distribution of X-ray generation, voltage/current fluctuations of X-ray generator, electronic noises in detector, beam geometry and divergence sampling, beam hardening, mechanical backlash.

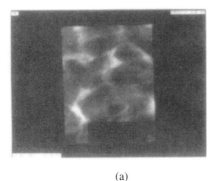
(a)

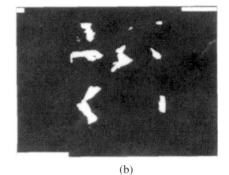
(b)

Fig. 3

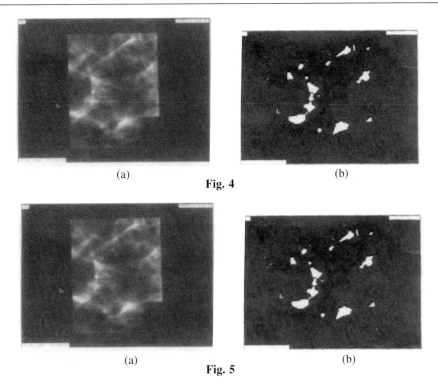

Fig. 4

Fig. 5

6. Conclusions

Literature relating to blast furnace operation and computerized tomography has been reviewed extensively. Typical X-ray radiography facility at NML has been created which will be the only facility available in the Eastern India region and an attempt has been made to develop open ended facility for the estimation of permeability distribution in a packed bed situation similar to blast furnace reactors at cold condition in the first phase through the processing of radiographic image slices in each scan. Post-reconstruction image processing may be beneficial to analyzing computerized tomography data if applied properly in this application. This powerful facility with high resolution can be used for qualitative as well as quantitative evaluation of thickness measurement, dimensional checks on complex variation, density variation, internal structure of complex components, design and manufacture, mapping of defective zones in components. This proposed system has wide range of applications in mines, nuclear, automobile, space and defence industries as well as in microtomographic applications.

References

1. Morgan, C.L., Basic Principles of Computed Tomography, 1st edn., University Park Press, Baltimore, MD, 1983.
2. A.C. Kak and Malcom Slaney, Computerized Tomographic Imaging, IEEE Press, 1988.
3. E. Cendre et al., 1st World Congress on Industrial Process Tomography, Buxton, Great Manchester, April 14-17, 1998.
4. Xu, B.H., Yu, A.B., Chew, S.J. and Zulli, Powder Tecnol., pp. 109-113, 2000.

5. ISIJ: Blast Furnace Phenomena and Modelling, Elsevier Applied Sci., London, pp. 376, 1987.
6. E1441, In: E-7; Vol. Radiology (X and Gamma) Method, ed. A.S.F.T.A. Materials, E07.01, Non-destructive Testing, p. 30, 1993.
7. Ravindran, V.R. et al., A Method for Evaluating the Radial Dimension of Flaws from Normal Radiographs of Large Propellant Grains, *British Journal of NDT*, **34**, pp. 294–296, 1992.
8. Bossl, R.H. and Netson, J.M, X-ray Computed Tomography Standards, Report WL-TR-94-4021, 1994.
9. Kumar, Umesh et al., Prototype Gamma-ray Computed Tomographic Imaging System for Industrial Applications, *INSIGHT*, **42**, 2000.
10. G.T. Herman, Image Reconstruction from Projections, Academic, New York, 1980.
11. X. Meng and Z. Hu, A New Easy Camera Calibration Technique Based on Circular Points, *Pattern Recognition*, **36**, No. 5, pp. 1155–1164, 2003.
12. Gonzalez, R.C. and Wintz, P., Digital Image Processing, Addison-Wesley, Reading, MA, 1987.
13. J. Zhang, X. Zhang, H. Krim and G.G. Walter, Object Representation and Recognition in Shape Spaces, *Pattern Recognition*, **36**, No. 5, pp. 1143–1154, 2003.
14. H.F. Lan and M.D. Levine, Finding a Small Number of Regions in an Image Using Low-level Features, *Pattern Recognition*, **35**, No. 11, pp. 2323–2340, 2002.
15. P. Sonol and P. Pendil, Feature Selection Toolbox, *Pattern Recognition*, **35**, No. 12, pp. 2749–2760, 2002.
16. Y.-L. Lee and R.-H. Park, A Surface-based Approach to 3-D Object Recognition using a Mean Field Annealing Neural Network, *Pattern Recognition*, **35**, No. 2, pp. 299–316, 2002.
17. A. Sanfeliu, R. Alquezar, J. Andrade, J. Climent, F. Serratosa and J. Verges, Graph-based Representation and Techniques for Image Processing and Image Analysis, *Pattern Recognition*, **35**, pp. 639–650, 2002.
18. G. Louverdis, M.I. Vardavoulia, I. Andreadis and Ph. Tsalides, A New Approach to Morphological Color Image Processing, *Pattern Recognition*, **35**, pp. 213–232, 2002.
19. Stanley, J.H., ASTM Standardization News, pp. 44–49, 1988.

Computerized Tomography for Scientists and Engineers
Edited by P. Munshi
Anamaya Publishers, New Delhi, India

8. Electrical Process Tomography: Imaging Fluid Mixing Processes Inside Stirred Vessels

R. Mann
School of Chemical Engineering and Analytical Science,
The University of Manchester, Manchester M60 1QD, UK

Abstract

Body-scanning exploiting 3-D imaging has revolutionised diagnostics and treatment in medicine. Process engineers would like to be similarly able to image chemical process units in 3-D, but without the £multi-million price tag. UMIST and Leeds University have together, through the Virtual Centre for Industrial Process Tomography (www.vcipt.org), pioneered several electrical process tomography techniques and used them in a variety of applications. Illustrations are presented to show how electrical resistance tomography (ERT) has been developed for typical stirred vessels widely encountered in batch process manufacturing. The technique is potentially fast and inexpensive and capable of imaging both dynamic and pseudo-stationary processes. Examples from UMIST's two-tonne vessel will be presented for miscible tracer mixing, as well as gas-liquid and solid-liquid mixing.

1. Introduction: Need for Better Instrumentation

Stirred vessels are widely used in the chemical industry as devices to provide fluid mixing. These vessels at manufacturing scale often have more than a passing resemblance to the beaker used in the chemical laboratory to gather information on chemical pathways and reaction kinetics. Such oversimplified scale-up, based on simple geometric similarity at the small and large scales, is supposed to greatly reduce the uncertainty in scaling-up for manufacture. This 'empirical' approach is forced upon us because of the general lack of understanding of the behaviour of such deceptively simple fluid mixing units. Although it is difficult to distinguish cause and effect in this scenario, there is little doubt that our lack of understanding interacts with a failure to provide instrumentation that can both measure and monitor the 'fluid mixing' on-line during typical batch manufacture.

Fluid mixing arises from complex interactions between the convective flow generated by the impeller/agitator rotation and the associated turbulence which has random eddies varying in size from the vessel dimension down to sub-micron. Turbulence is invariably present for low viscosity thin water-like fluids, but even at higher viscosities, a thicker, more viscous fluid will show complications instead from non-Newtonian rheology, even though turbulence may be absent.

In chemicals manufacture, the effects of fluid mixing on the chemical pathway(s), hence the selectivity to desired product(s), is always indirect and arises via chemical species concentrations. Whenever the reactions are fast relative to rates of fluid mixing, there will always be a spatial 'field' of concentrations under semi-batch operation. In this mode of operation, one chemical component is added from a feed reservoir to another already in the vessel. In this way, the rationing of one of the reagents controls reaction run-away hazards. However, this drip feed of one reagent when combined with the usual need to have reasonable production rates will invariably lead to non-uniform internal concentration fields. This in turn may have a complex effect upon the chemistry and spectrum of chemical products. It is therefore highly desirable to be able to quantify such concentration fields, which will always be three-dimensional for single point addition.

Unfortunately, our capabilities for 'visualising' 3-D concentration fields inside a stirred vessel were until recently non-existent. Certainly, the classic approach using stimulus-response techniques (Levenspiel, 1999) is of no use for batch stirred vessels since they have no through-flow! The alternative approach seeks to use the mixing time, which is the time for a pulse injected material to become homogenised to some specific degree. However, as this is only a simple scalar measure of the rates of fluid mixing, it is of limited use for understanding the complex interactions of fluid mixing and chemistry.

This article describes some recent developments in electrical resistance tomography (ERT) which have enabled us to begin to visualise mixing processes in 3-D inside a typical stirred vessel. ERT is inexpensive when compared to X-ray and MRI techniques developed for medical imaging. With ERT, the £-million price tag for a body-scanner is reduced to a few £-thousand. In any case, this much cheaper technique is dominated by the cost of the associated electronics, since the sensor hardware can be made from simple metal plates wired together. The technique is effectively both non-intrusive and non-invasive and hence does not interfere with the internal fluid mechanics.

2. Theoretical Insights Confirming Internal Spatial Non-uniformity

Evidence for the existence of spatially distributed concentration fields has been accumulating for a number of years, earlier theoretical analyses being limited to 2-D axisymmetry to reduce the computational burden (Mann, 1995). More recently, it has proved possible to compute in 3-D and some results are presented in Fig. 1 for single phase miscible fluid mixing with single point addition of a component B from a feed reservoir to a component A initially placed in the vessel (Rahimi and Mann, 2001). Because there are two possible reactions of A, these predictions could also include chemical yield/selectivity. These computations use a simplified fluid dynamics based on networks of backmixed zones, which incorporates an overall rate of internal convective flow generated by the impeller rotation, a turbulent mixing parameter and a convective swirl (tangential) flow magnitude. In Fig. 1, visual reality images are created based upon colouring each voxel according to its component concentration and ascribing it an optical opacity also based upon concentration. What Fig. 1 confirms is that there is predicted to be a distinctive plume of A (shown in red) spreading out from the addition point. By making use of a combination of forward perspective and overhead plan views, it is easy to visualise how the plume enlarges and swirls tangentially as the addition of B proceeds linearly over 15 s. The use of variable opacity allows an element of see-through, so that as the reagents disappear by reaction, the fluid becomes clearer (less opaque), and thus by 30 s, which is 15 s after completing the addition, the impeller has become partly visible. Predictions such as these need a visualisation technique like ERT, in order to validate the theory.

Similar effects can be expected for reactions which involve multiple phases and mass transfer. Thus, Fig. 2 shows an image of predicted two-phase gas-liquid mixing in which bubbles fed through a sparge ring will be dispersed and distributed by the radial flow pattern typically produced by a flat-blade or Rushton turbine (Williams et al, 1997). The bubbles are not uniformly dispersed throughout the liquid phase since they rise differently in upflow, downflow and crossflow. In particular, there are few bubbles in the lower part of the vessel, and a concentration of them just above the sparge injection point. The pattern here is axisymmetric. It is evident that ERT,

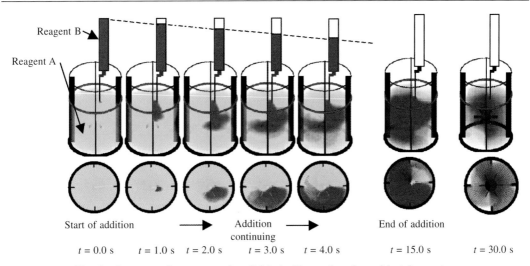

Fig. 1 Segregated concentration fields inside a stirred semi-batch reactor.

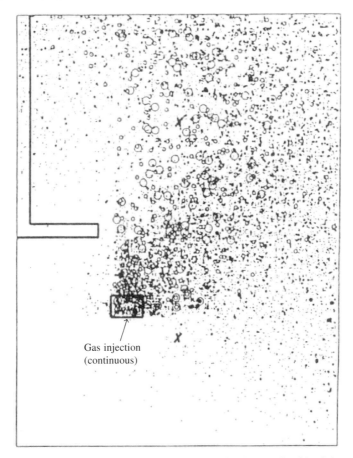

Fig. 2 Non-uniform gas hold-up distribution in gas-liquid mixing.

because of its capability to measure the local electrical conductivity, should be able to detect and quantify this gas-voidage/hold-up field. If the gas and liquid each provide chemical components that can react, then the complex interactions of two-phase mixing and interphase mass transfer could give rise to regions rich in each component as shown schematically in Fig. 3. Some

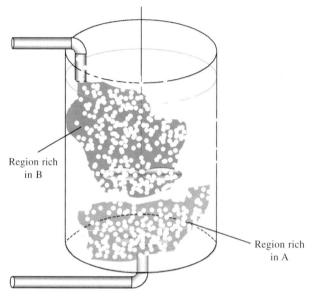

Fig. 3 Segregated concentration fields for a gas-liquid reactor.

predictions in 3-D have recently shown how dissolved oxygen and nutrient can be partially segregated in this way in a stirred bioreactor (Hristov et al, 2001; Hristov et al, 2004). Fig. 4

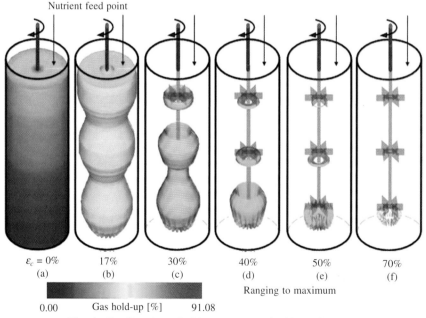

Fig. 4 Predicted gas hold-up contours inside a bioreactor.

shows some predicted gas hold-up contours for a bioreactor fitted with three radial-flow Rushton turbines (Hristov et al, 2001). Once again, ERT can in principle visualise this type of behaviour, although now the technique needs to be able to be chemical species specific if the chemical species segregation is to be visualised in accordance with Fig. 3. Nevertheless, these three practical examples show the kind of internal visualisation needs of typical stirred vessels as used in the chemical industry.

3. A Practical ERT System for a Plant-scale Stirred Vessel

A schematic view of the system installed on UMIST's two-tonne pilot plant vessel is shown in Fig. 5. The hardware forming the set of sensors is made up of 16 simple equally spaced rectangular stainless plates formed into a circular ring. There are eight such equally axially spaced rings as seen in the left part of Fig. 5. The so-called interrogation protocol, in which current is injected and detected for the full set of combinations of injection/detection for each ring is set and executed by the data-acquisition system run by the PC. Signals returned from the DAS are passed to the PC, where they can be reconstructed into a tomogram for each of the eight circular planes. This system has been described in more detail elsewhere (Mann et al, 1997).

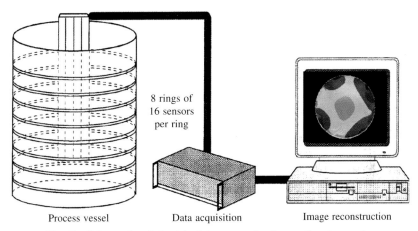

Fig. 5 Schematic of electrical tomography for a stirred vessel.

A view of the vessel equipped in this way is given in Fig. 6. The vessel's Rushton impeller in a standard configuration can be clearly seen, as can the eight rings of sensing planes. The stainless steel plates appear as dark grey rectangles against the wall-mounted white trunking which holds them. The cylindrical vessel is filled with water to a depth of 1.45 m, exactly equal to the vessel diameter (the so-called square configuration).

Fig. 7 shows schematically how the current injection and voltage detection works using the adjacent pairs protocol. For each injection, the resistivity along the 'path' between equipotential lines is registered by the measured voltage. Then, as shown in Fig. 7(b), for two measurements, the registered voltages contain joint information where they overlap. For a large number of measurements restricted to the outer periphery, the whole field can be covered and the interior values "reconstructed" from the set of peripheral values in the classic tomographic analysis. In this electrical process tomography, the interrogating paths are inconveniently non-straight, unlike X-rays. Reconstruction from such non-linear so-called "soft" fields is then less accurate, although simplified techniques like the simplest linear back-projection may still be capable of useful

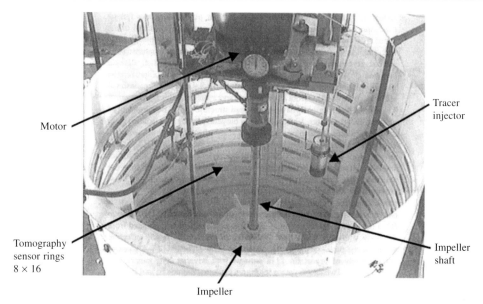

Fig. 6 View of two-tonne stirred vessel in UMIST pilot-plant.

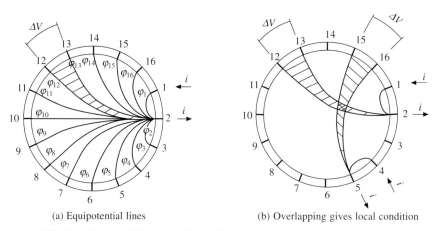

(a) Equipotential lines (b) Overlapping gives local condition

Fig. 7 Current injection/voltage detection for adjacent pairs protocol.

imaging of many processes. Finally, although this system neglects out of plane electrical effects, the set of eight tomograms can be "stitched" together to create a pseudo 3-D image.

4. Stirred Vessel Tomographic Images in 3-D: Some Examples

4.1 Miscible Tracer Mixing

In this section, some illustrative results from the two-tonne vessel are presented in brief. For each example, more complete details have been published elsewhere.

Figure 8 shows how the details of miscible fluid mixing can be captured over a few seconds, using a pulse injection of a highly conducting fluid (Holden et al, 1998; Mann et al, 2001). The point of injection of a brine tracer pulse is shown in the left hand pair of images at $t = 0$. As before, the tomographic reconstructions are shown in pairs of images, with a forward perspective view on top and an overhead plan view below. The views at $t = 0$ are blank. The ensuing electrical conductivity fields in time progression are then visualised as partial see-through assemblies of sets of solid-body contours coloured according to concentration. The intense high concentration red colour shrinks and fades as dilution takes place in three dimensions by fluid mixing. The green outermost boundary contour correspondingly enlarges as mixing proceeds, and the views show how the salt mixes through the impeller and undergoes clockwise swirling under the rotation of the impeller. It is important to recognise that although these images look like typical computational fluid dynamics (CFD), they are in fact experimental results. The 16-sensor × 8-ring array reconstructs 104 pixels for each of the eight planes, which in turn creates 832 pseudo voxels for the 3-D representation. Thus, each image in Fig. 8 contains $O(10^3)$ experimental points. Since the ERT system can acquire an image in about 0.3 sec, 1 Mb of mixing information is furnished in each 3 s.

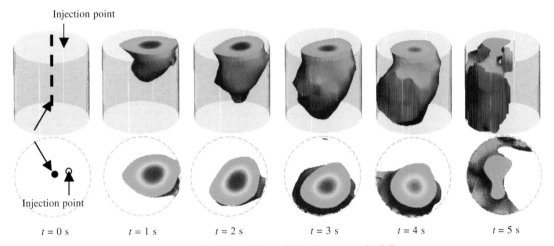

Fig. 8 Miscible mixing of a brine tracer in 3-D.

4.2 Gas-liquid Mixing

Two example images are presented showing how ERT can be used to visualise two-phase gas-liquid mixing. Fig. 9 shows an experimental result for gas hold-up variation presented as three iso-surface contours. This image is for the type of situation predicted in Fig. 2. Here regions of high/low conductivity can be detected, but not the individual bubbles. It is also not possible to say anything about variations in bubble size due to the limited spatial resolution of our ERT set-up. Fig. 9 shows that the gas hold-up contours are approximately axisymmetric, but there is some small random variation in the expected symmetry caused almost certainly by the "noisy" turbulent flow typical of such gas-liquid stirring.

Fig. 10 shows an originally unexpected capability of this ERT system to monitor mixing in just the continuous liquid phase. This can be readily achieved in practice by adding a pulse of brine tracer. The large changes in liquid conductivity can be separately imaged if the original gas-liquid voidage field is used as a new background baseline. The results in Fig. 10 are shown in this instance as stacks of eight tomograms. These clearly show in each case how the fluid progresses towards homogeneity as the tracer is subjected to fluid mixing when gas-liquid

Fig. 9 Solid-body contours of gas hold-up for gas-liquid mixing.

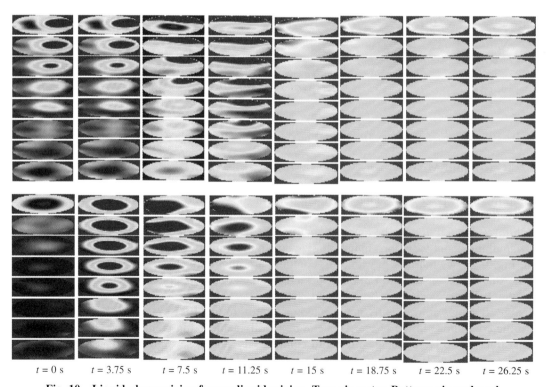

Fig. 10 Liquid phase mixing for gas-liquid mixing. Top: air-water, Bottom: air-carbopol.

mixing is simultaneously taking place. The two cases in Fig. 10 refer to the effect of difference in liquid viscosity. The upper set is for viscous carbopol-air and the lower one for much less viscous water. These results clearly show the difference in mixing rates in the liquid for the two cases.

4.3 Solid-liquid Mixing

Finally, two examples of application to two-phase liquid-solid mixing are presented. In processes where the solid and liquid have differing electrical conductivities, ERT has the potential to image the distribution of solid when stirred with a liquid. This is frequently undertaken in manufacturing using stirred vessels. In this case we have imaged the suspension of non-conducting polypropylene particles of mm dimensions at low (1%) solid loadings. The nature of the behaviour is shown for illustration in Fig. 11 using conventional visualisation in a lab scale glass stirred vessel filled with (clear) water. In Fig. 11 the particles are black. Being denser than water they tend to settle on the vessel base unless the stirring is sufficient to keep them fully in suspension. The left hand image shows a forward perspective view together with an underneath view gathered by placing a mirror beneath the vessel at 45°. At the stirrer speed of 100 rpm, almost all the particles are formed into a symmetrical 'hill' with some particles lifted up towards the impeller. Interestingly, the underneath view shows that the particles are not uniformly distributed across the base, but form heaps around the baffles, with some portions of the base swept clean of particles. At a higher speed of 150 rpm in the right hand view, particles are better suspended and reach above the impeller. The base is still not uniform in deposited particles, although the self-obscuration by the suspended particles now makes it more difficult to see the clear portions of the base. However, at both stirrer speeds, it is clear that the four vertical baffles introduce a 3-D effect.

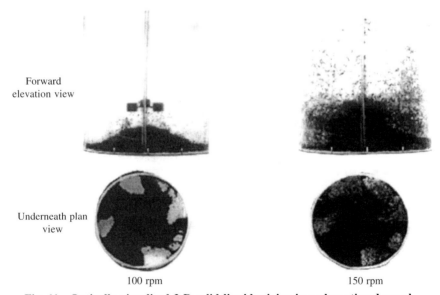

Fig. 11 **Optically visualised 3-D solid-liquid mixing in a glass stirred vessel.**

Fig. 12 shows how behaviour equivalent to normal visualisation in Fig. 11 can be captured by ERT and image-reconstructed into 3-D images (Mann et al, 2001). These results are from the two-tonne vessel with 1% (?) of the same solid polypropylene particles. In this visualisation, we have retained a perspective forward view and an overhead plan view as before, but in this case interposed these with the forward elevation view, thus providing a set of three images at each pseudo-stationary solid-liquid mixing condition. Now, however, by converting the voxel conductivity values into solid content, as the particles have zero conductivity, we are able to map the solid distribution. Then, a set of three solid-content contours have been interpolated from the $O(10^3)$ voxels in Fig. 12. Between each contour an increasing degree of opacity has been ascribed to the intervening solid-content level, so that the highest level is fully opaque and below the lowest contour is perfectly clear. When the figure is constituted in this way, an augmented-reality image is generated which visually conveys the nature of internal solid suspension created by the impeller action. Thus, in Fig. 12, the darkest volumes belong to the highest solid-content, which is shown to lie behind the baffles on the downstream side (with clockwise impeller rotation). Fig. 12 shows that as the impeller speed is increased, these highest density 'pockets' of solid diminish in size. This is then balanced by the larger liquid volume at the intermediate solid level. This coincides with solid suspended at increasing heights above the base. These results therefore capture the full 3-D character of solid suspension in a way never achieved before. Moreover, it is important to stress that these ERT visualisations could be achieved even if the fluid in which the solids were being suspended was actually totally opaque, so nothing could be seen by the naked eye. It can therefore be reasonably claimed that Fig. 12 represents seeing inside a stirred vessel 'without eyes'.

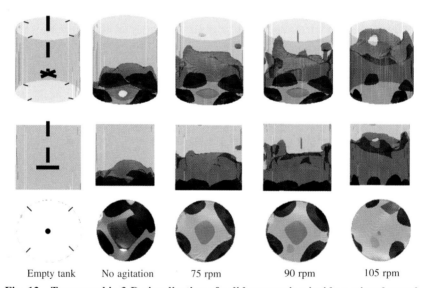

Fig. 12 Tomographic 3-D visualisation of solid suspension inside a stirred vessel.

Finally, on an issue related to solid-liquid mixing, Fig. 13 shows recent results for visualising semi-batch operation (Stanley et al, 2002). The arrangement is similar to that in Fig. 1, but now if the reaction between the added reagent and the one already in the stirred vessel resulted in the precipitation of a solid product, it would be necessary to understand the composition field of the evolving plume. It has long been suspected that the "character" of the solid particles which

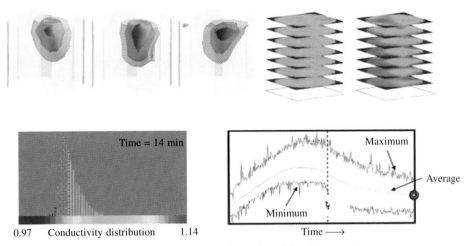

Fig. 13   **Tomographic imaging of semi-batch operation.**

nucleate and grow in the vessel will be sensitively dependent upon the local concentrations, supersaturation level and electrical conductivity. The physico-chemical fundamentals are, however, so complicated that our basic understanding is hopelessly insufficient to control the resulting particle sizes and morphology. This is especially needed in the pharmaceutical industry, where the solid product ought to possess a set of desirable properties that will engender good efficacy as a (often highly active and expensive) drug. Fig. 13 shows that we are able to visualise the addition plume in 3-D as semi-batch operation progresses. Also shown in Fig. 13 are the voxel distributions produced by ITS's software (www.itoms.com), as well as the time-wise traces of the max, mean and min voxel values (which encompass the entire variance of electrical conductivity within which the semi-batch fluid mixing addition is confined). It is hoped again that the augmented-reality imaging of the semi-batch plume shown in Fig. 13 will provide valuable assistance in understanding and quantifying the complexities of fluid mixing accompanying semi-batch precipitation.

5.   Conclusions

(a) Electrical resistance tomography provides a powerful means for non-intrusive measurement of fluid mixing processes inside typical stirred vessels.
(b) Variations in conductivity enable 3-D visualisation of miscible tracer mixing, as well as gas-liquid and solid-liquid mixing.
(c) Solid-body graphics with opacity variation can create augmented-reality 3-D imaging.
(d) Emerging developments will provide simpler techniques for better process imaging and reality augmentation.

Acknowledgements

I am thankful to my research students Zahira Yaakob, Dimiter Vlaev, Masoud Rahimi, Hristo Hristov, Eustace Wabo and Steven Stanley who carried out the work covered in this article; EPSRC for Grant GR/J50798 and for a Total Technology studentship; Barry Edwards and Unilever for additional financial support and finally to ITS Ltd.

References

Holden, P.J., Wang, M., Mann, R., Dickin, F.J. and Edwards, R.B. (1998) "Imaging stirred vessel macromixing using electrical resistance tomography" *A.I.Ch.E.J.*, **44**, 780–789.

Hristov, H., Mann, R., Lossev, V., Vlaev, S.D. and Seichter, P. (2001) "A 3D analysis of gas-liquid mixing, mass transfer and bioreaction in a stirred bio-reactor", *Trans. I. Chem. E.*, **79(C)**, 232–239.

Hristov, H., Mann, R., Lossev, V. and Vlaev, S.D. (2004) "A simplified CFD for three-dimensional analysis of fluid mixing, mass transfer and bioreaction in a fermenter equipped with triple novel geometry impellers", *Trans. I. Chem. E.*, **82(C1),** 21–34.

Levenspiel, O. (1999) "Chemical Reaction Engineering", 3rd Edition, Wiley, New York.

Mann, R., Pillai, S.K., El-Hamouz, A.M., Ying, P., Togatorop, A. and Edwards, R.B. (1995) "Computational fluid mixing in stirred vessels: Progress from seeing to believing", *Chem. Eng. Jl.*, **59**, 39–50.

Mann, R., Dickin, F.J., Wang, M., Dyakowski, T., Williams, R.A., Edwards, R.B., Forrest, A.E. and Holden, P.J. (1997) "Application of electrical resistance tomography to interrogate mixing processes at plant scale", *Chem. Eng. Sci.*, **52**, 2087–2097.

Mann, R., Wang, M., Forrest, A.E., Holden, P.J., Dickin, F.J., Dyakowski, T. and Edwards, R.B. (1999) "Gas-liquid and miscible liquid mixing in a plant-scale vessel monitored using electrical resistance tomography", *Chem. Eng. Commun.*, **175**, 39–48.

Mann, R., Stanley, S.J., Vlaev, D., Wabo, E. and Primrose, K. (2001) "Augmented-reality visualisation of fluid mixing in stirred chemical reactors using electrical resistance tomography (ERT)", *J. Electronic Imaging*, **10(3)**, 620–629.

Rahimi, M. and Mann, R. (2001) "Macro-mixing, partial segregation and 3D selectivity fields inside a semi-batch stirred reactor", *Chem. Eng. Sci.*, **56**, 763–771.

Stanley, S.J., Mann, R. and Primrose, K. (2002) "Tomographic imaging of fluid mixing in 3-D for single-feed semi-batch operation of a stirred vessel, *Trans. I. Chem. E.*, **80(A)**, 903–909.

Wang, M., Dorward, A., Vlaev, D.S. and Mann, R. (2000) "Measurements of gas-liquid mixing in a stirred vessel using electrical resistance tomography", *Chem. Eng. Jl.*, **77**, 93–98.

Williams, R.A., Mann, R., Dyakowski, T., Dickin, F.J. and Edwards, R.B. (1997) "Development of mixing models using electrical resistance tomography", *Chem. Eng. Sci.*, **52**, 2073–2085.

Computerized Tomography for Scientists and Engineers
Edited by P. Munshi
2006, Anamaya Publishers, New Delhi, India

9. Convection in Differentially Heated Fluid Layers and Its Reconstruction Using Radial Tomography in an Octagonal Cavity

Sunil Punjabi
Department of Mechanical Engineering, Ujjain Engineering College, Ujjain-456 010, India

Abstract

Buoyancy-driven convection in differentially heated fluid layers confined in an octagonal cavity has been experimentally studied using interferometric tomography. Combinations of air and water, and air and silicone oil have been considered. The grade of silicone oil employed in the experiments is 50 cSt. Fluid layers are confined between two horizontal isothermal plates and the enclosure is heated from below and cooled from above. The thermal configuration leads to a modified form of the Rayleigh-Benard problem that has been widely studied in the literature for theoretical as well as practical reasons. The flow field was mapped from four different view angles, namely 0, 45, 90 and 135° using a Mach-Zehnder interferometer for each fluid combination and cavity based temperature difference. The projection data obtained from different view angles was utilized in reconstructing the three-dimensional temperature field using a tomographic algorithm. Convolution back projection (CBP) classifies as a transform method has been chosen as the tomographic algorithm. Three-dimensional reconstruction at three selected planes in the fluid layers confirmed the thermal field to be axisymmetric, silicone oil in particular. A plume structure was identified in the flow field on the basis of reconstruction.

1. Introduction

Convective heat transfer refers to energy transport in a fluid medium, enhanced by the fluid velocity. In natural convection, the fluid motion is set up solely by the buoyancy forces arising from the presence of density gradients. Changes in density can be induced by thermal fields, concentration fields or a combination of the two. The resulting flow pattern will additionally depend on the orientation of the density gradient with respect to the gravity field, the geometry of the confining boundaries and boundary conditions.

The present work investigates the buoyancy-driven convection in two superposed fluid layers. The experimental cavity is octagonal in-plan that closely approximates to a circular cavity. The thermal field has been mapped using a Mach-Zehnder interferometer. The interferograms have been obtained as a collection of fringe patterns. The interferograms give line-averaged information along the direction of the light ray. Hence, they represent a projection of the thermal field in that direction. The present cavity allows the projection data of the thermal field from four different view angles, namely 0, 45, 90 and 135°, while the fifth angle 180° is similar to 0° view angle. The positions of the source (laser) and the detector (CCD camera) are held constant during the experiments. For reconstruction of the thermal field, the transform algorithm namely convolution back projection (CBP) has been adopted. The two fluid layer combinations selected for the tomographic reconstructions of the temperature field were air-water and air-silicone oil. The grade of silicone oil employed in the experiments was 50 cSt. The fluid layer thicknesses were maintained equal for each combinations selected.

The objectives of this article are to examine the nature of the axisymmetric thermal fields in the fluid layers, the influence of increasing Rayleigh number on transition to three-dimensionality, and the appearance of unsteadiness. These objectives have also been examined in terms of the reconstructed three-dimensional temperature field from its interferometric projections using principles of tomography.

Important applications of flow and heat transfer in superposed fluid layers can be found in encapsulated crystal growth from its melt, growth of optical crystals from the aqueous solution, and buoyancy-driven convection in stratified water bodies such as solar pond. The data from large aspect ratio, high Prandtl number experiments can be applied to study convection in the earth's mantle.

Two-layer convection has been studied in cylindrical, rectangular and square geometries by several authors. In the studies reported, the focus has been on the destabilizing effect of surface tension gradients at the fluid-fluid interface on the onset of buoyancy-driven convection. In contrast, the size of the cavity and the depth of the fluid layers of the present experiments are such that surface tension effects are of secondary importance. Experiments have been conducted in post-stability regime of buoyant convection in the fluid layers. The cylindrical geometry has been adopted in the present work because of its application to crystal growth.

A linear stability analysis of buoyant convection in superposed fluid layers has been presented in [1]. The layer thicknesses were such that Marangoni effects were generally small. The interface deformation was included in the analysis, and a wide range of parameters was investigated. Conditions under which the onset of convection had viscous or thermal coupling, and exhibited Hopf bifurcation were determined. The theoretical predictions were compared with experiments carried out in a rectangular cavity whose length was much greater than the cross-sectional dimensions. The fluids chosen were combinations of ethylene glycol-oil and ethylene glycol-decane. The measurement was based on the shadowgraph technique, with signals being recorded by a photodiode. The mode shapes realized in the experiments matched theory, though oscillatory convection and periodic switching of coupling modes were not experimentally verified. Pattern selection at the onset of instability in small aspect ratio containers with a free surface has been analyzed in [2]. The geometry considered was such that interface deformations were small, but in addition to buoyancy effects, Marangoni convection was significant. The analysis showed that the no-slip side walls greatly influenced the number of multiple stable patterns possible in comparison to free slip boundaries. Double-layer convection has been discussed from a theoretical and experimental viewpoint in [3], in the context of geophysical applications. Experimental results for two immiscible fluid layers, driven by a vertical temperature difference slightly greater than critical for Benard instabilities have been reported in [4]. Both rectangular and annular cavities were considered. The condition under which the onset of convection is time dependent was seen to be a function of the property ratio of the liquids constituting the fluid layers. A theory of secondary instabilities in liquid-gas system with a deformable interface was proposed in [5]. The flow field was taken to be driven exclusively by Marangoni convection, buoyancy effects being negligible because of the small layer heights. A wide variety of short- and long-scale patterns were seen to be possible, in addition to the excitation of deformational waves at the interface. Experiments and numerical simulation for flow and thermal fields in the post-stability regime have been reported for rectangular cavities in [6-7].

In the context of crystal growth, two-layer convection has been studied in a cylindrical geometry by various authors. During liquid encapsulated crystal growth, the layer aspect ratio changes with time due to depletion of the melt. The effect of lateral boundaries on the flow pattern that arises when interfacial- and gravity-driven convections are present is an interesting aspect of two-layer convection. The effect of the container geometry on the flow pattern that forms near the onset of interfacial-driven convection has been studied in [8]. The numerical studies were carried out in a cylindrical geometry with two radius-to-height ratios of 1.5 and 2.5. The critical Marangoni numbers were calculated from the linearized Boussinesq equations, the interface deformations being assumed to be negligible. In continuation, the authors conducted experiments to study the effect of the layer height on the pattern formation in silicone oil-air system [9]. For flow visualization an inframetric camera was used. The camera measured the infrared radiation emitted by top surface of the silicone oil and thus the temperature distribution at the interface. The convection pattern set up in air was found to be quite weak to initiate mechanical coupling. However, the thermal gradients were setup along the interface, leading to convection in the silicone oil layer due to buoyancy as well as surface tension gradients.

Application of computerized tomography for measurements in heat and mass transfer has been a topic of research over the past decade. Many authors have reported the three-dimensional reconstruction of thermal fields using the tomographic algorithms. Different optical measurements techniques and application of tomographic algorithms to evaluate three-dimensional fields has been reviewed in [10].

The three-dimensional temperature field was reconstructed in [11] in a differentially heated fluid layer of the Rayleigh-Benard problem from its interferometric projections. The authors saw the presence of rolls in a water filled cavity of aspect ratios 8.7 and 9.0. The two horizontal confining walls of the cavity were made of aluminium. Two sides of the vertical side walls were made from delrin and the other two sides were made from 25 mm thick optical flats. Three thermistors were used in each aluminium plate to measure the temperature of the plates. A Mach-Zehnder interferometer with 20 cm diameter optics and a helium-neon laser of 10 mW power were employed for collecting the projection data. The wedge fringe setting of the interferometer was used to record the convection pattern inside the cavity. For the tomographic reconstruction of the three dimensional temperature field the authors employed the corresponding numerical solution of the physical problem as an initial guess to the reconstruction algorithm. The results showed the formation of T-defects in the longitudinal rolls near the region of the short side of the box. The movement of hot and cold spots along the roll was observed.

Mishra et al [12-14] reported an experimental study of Rayleigh-Benard convection in an intermediate aspect ratio box that was square in plan. An intermediate range of Rayleigh numbers namely 13900, 34800 and 40200 was considered in the study. The fluid employed was air. A Mach-Zehnder interferometer was used to collect the line-of-sight projections of the temperature field in the form of interferometric fringes. Interferograms were collected from four to six view angles. These were used to obtain the three-dimensional temperature field inside the cavity by using tomography. An algebraic reconstruction technique (ART) was used for the inversion of the projection data. The convergence of the iterative inversion procedure was unambiguous and asymptotic. The reconstructed temperature field with a subset of the total data was found to be consistent with the remaining unused projections.

In the present work, buoyancy-driven convection in differentially heated two superposed fluid layers enclosed in an octagonal cavity has been experimentally studied using interferometric tomography. The fluid layers are horizontal with respect to the direction of the gravity field. The projection data obtained in the form of interferograms from the experiments are utilized for the recovery of local temperature using the principles of tomography. The thermophysical properties for air, water and silicone oil are given in Table 1 at 25°C, the ambient temperature maintained during the experiments. The temperature differences applied across the cavity were large enough for the Rayleigh numbers to be sufficiently greater than the respective critical values. For the cavity diameter and layer thicknesses of the present study, interface deformation owing to surface tension as well as Marangoni convection were estimated to be of secondary importance.

Table 1. Thermophysical properties of air, water and silicone oil at 25°C

Fluid	ρ (kg/m^3)	ν (m^2/s)	μ (kg/m/s)	K (W/m-K)	α (m^2/s)	β (K^{-1})	Pr
Air	1.184	15.5E-06	18.4E-06	26.1E-03	2.19E-05	3.35E-03	0.71
Water	997.05	89.3E-08	89.04E-05	0.611	1.46E-07	20.57E-05	6.1
Silicone oil	960	5.00E-05	4.80E-02	0.151	1.05E-07	1.06E-03	476.2

The convection patterns in the cavity have been imaged using a laser interferometer. The light beam orientation resulted in the depth-averaging of the respective thermal fields in the fluid layers. The objectives of the present work are to examine the influence of Rayleigh number on the steady thermal field, temperature field developed in the individual fluid layers and the tomographic reconstruction of the temperature contours at three selected planes using a transform algorithm namely convolution back projection (CBP).

2. Apparatus and Instrumentation

Rayleigh-Benard convection forms a subclass of applications involving buoyancy-driven flows and thermal convection. The experiments have to be conducted with due care and precaution owing to complexities seen in the flow patterns that form in the fluid medium. Further, the quality of experiments are determined largely on the basis of uniformity and constancy of horizontal surface temperatures, parallelism of the walls defining the fluid layer, parallelism between optical windows placed at the entrance and exit to the laser beam and properties of the insulating side wall surfaces.

The test cell used to study buoyant convection in superposed fluid layers is shown in Fig. 1. It consists of three sections, viz. top tank, test section (cavity) and the bottom tank. The test section is octagonal in plan and has a nominal diameter of 130.6 mm and a height of 50 mm. The cavity aspect ratio defined as the diameter-to-height ratio is 2.61; for the fluid layers it is 5.22 for all the experiments. These values categorize the cavity as a small aspect ratio container [15]. The fluid layers, 25 mm thick are confined by two copper plates of thickness 1.6 mm above and below. The test cavity is essentially made of optical windows, 50 mm square and 3 mm thick, with 8 of them forming the octagon in plan. The windows are essential in the present work since they allow viewing of the thermal fields at parallel incidence and at various angles. For the octagonal geometry adopted for the experimental apparatus, view angles of 0, 45, 90 and 135° are possible. The optical windows are made of commercially available laser-grade fused silica. The high quality windows permit the passage of the laser beam without refraction. The hot and the cold surfaces were maintained at uniform temperatures by circulating water from constant temperature baths (RAAGA and HUBER). A tortuous flow path is created in the lower tank by installing 5 baffles that additionally act as fins for the lower surface. The increased contact area enhances heat transfer and distributes temperature uniformly over the lower plate. For the top tank, contact area between the flowing water and the copper surface was readily available. Both

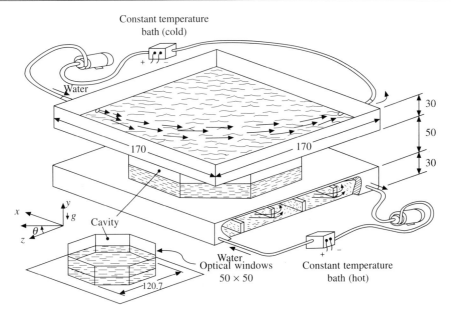

Fig. 1 Schematic drawing of an octagonal (nominally circular) test cell to study convection in superposed fluid layers (all dimensions are in mm).

horizontal surfaces have been maintained at their respective temperatures to within ±0.1 K during the experiments. The surface areas of two tanks have been designed to be much larger than the cavity to reduce edge effects. All experiments continued for 4 h to confirm that steady state was reached in the fluid layers. Although the flow field evolved fully in 2 h, the experiment was continued for an additional 2 h for detecting mild changes in the fringe field. The time required for the thermal fields in the hot and cold surfaces was found to be of the order of just a few minutes. In this respect, the thermal loading of the two-layer system may be categorized as a step change.

To resolve near wall fringes, the two copper plates were carefully checked for flatness and surface finish. Flatness of the plates registered against a reference face plate was around 50 µm. The surface finish measured by an instrument (SURFANALYZER 5000) showed the RMS value to be close to 0.4 µm. During the experiments, the optical windows were additionally covered by 12 mm thick plexiglas in order to insulate the test section with respect to the atmosphere. The plexiglas blocks of the opposite pair of optical windows were removed during data collection.

For measurements in liquids a reference chamber is required to be included with the interferometer to compensate for refractive index changes under isothermal conditions. The reference cell utilized in the present experiments is rectangular in construction. It is placed in the compensation chamber of the interferometer and is thermally inactive. By including the reference chamber, the interferograms reveal exclusively the variations in the thermal field in the cavity.

For temperature field measurement in the fluid medium, a Mach-Zehnder interferometer has been employed. It uses a 35 mW He-Ne laser (SPECTRA PHYSICS) and 150 mm diameter optics. Interferograms have been recorded using a CCD camera (PULNIX) that has a 512×512 pixel resolution. The camera is interfaced to an IBM-compatible PC through an 8-bit A/D card which digitizes the light intensity levels over the range of 0-255. Image acquisition is at video rates (50 images/s). Temperatures at various points over the solid surfaces of the cavity as well as the ambient temperature have been monitored by 18 gage K-type thermocouples. The thermocouples are in turn connected to a 30 channel recorder (SAN-EI) for the duration of the experiment.

3. Tomography

The three-dimensional temperature field can be reconstructed from its interferometric projections using principles of tomography. Tomography is the process of recovery of a function from a set of its line integrals evaluated along some well-defined directions. In interferometry, the source of light (the laser) and the detector (CCD camera) lie on a straight line with a test-cell in between. Further, a parallel beam of light is used. This configuration is called *transmission tomography* in a *parallel beam geometry* [16]. Tomographic algorithms used in interferometry reconstruct two-dimensional fields from their one-dimensional projections. Reconstruction is then applied sequentially from one plane to the next till the third dimension is filled.

Tomography can be classified as: (a) transform, (b) series expansion and (c) optimization methods. Transform methods generally require a large number of projections for a meaningful answer. In practice, projections can be recorded either by turning the experimental set-up or the source-detector combination. In interferometry, the latter is particularly difficult and more so with the Mach-Zehnder configuration. With the first option, it is not possible to record a large number of projections, partly owing to inconvenience and partly due to time and cost. The limited projection data available from interferometry can, however, be suitably increased for implementing the transform method for reconstruction.

In measurements involving commercial grade optical components and recording and digitizing elements, the projection data is invariably superimposed with noise. Software operations such as interpolation and image processing can also contribute to errors in the projection data. Experience of the author in [12] with interferometric experiments shows that the RMS noise level of greater than 5% can result in unphysical artifacts in the solution, if iterative methods are used.

In the present work, the transform method is adopted for reconstruction of three-dimensional temperature field in an octagonal cavity. Convolution back projection (CBP) has been chosen as the tomographic algorithm in view of its established convergence properties. In situations when measurements for numerous projections over the total viewing angle of 180° are possible, the back projection method is recommended. These methods are usually faster to compute and require less computer memory than those using series expansion [10].

With the present experimental set-up, only four view angles were possible for generating the projection data, viz. 0, 45, 90 and 135°. The projection data obtained at a view angle of 180° (fifth view) is identical to that of 0° view angle. For each view angle, the size of the light beam covered 41% of the full width of the fluid layer in the cavity. This corresponds to a partial projection data that is captured in the central portion of the cavity.

In tomography, the projection data required for reconstruction should be available for the full width of the flow field from each view angle rather than the partial width, that is, the light beam should scan the thermal field covering the entire width of the cavity for all view angles. In the present experimental configuration, it was not possible to scan the full width owing to the laser beam diameter (72 mm) being less than the total nominal diameter (130.65 mm) of the cavity. To overcome this limitation, the projection data was increased by applying suitable interpolation and extrapolation schemes in which the number of view angles were increased from 5 to 81 by linearly interpolating the experimental data available for the central portion. This step does not increase the information content of the projections. The obtained interpolated data still covers the partial width of the fluid layer. In order to scan the full width of the thermal field, a suitable extrapolation procedure was adopted. This extrapolation scheme was implemented with reference to a circle that closely approximates the shape of the axisymmetric test cell. For a circular geometry, the chord length of integration diminishes from the diameter of the circle at the center to zero at its edges. Thus, the sensitivity of the interferometric measurement is at its highest closer to the center, as compared to the sides. The drop in sensitivity is seen in the form of a drastic reduction in the number of fringes. Inverting this data to recover temperature is mathematically inappropriate. Tomographic algorithms such as CBP require projection data for the complete width of the fluid layer. The extrapolation procedure adopted in the present work comprises assuming the temperature field to be spatially uniform outside the central core of 41%. Allowances for the change in the chord length are, however, included. This procedure is not expected to affect the quality of the original partial projection data obtained directly from the experiments.

The projection data for each view angle should be consistent in terms of the average fluid temperature over a given plane of the fluid layer. In other words, the average temperature is constant for a given plane irrespective of the view angles. This check was enforced in the line integral data before reconstruction.

The salient features of the CBP algorithm for reconstruction, as discussed by [16] and in a review article [17] have been reviewed in the following section.

3.1 Convolution Back Projection

The convolution back projection (CBP) algorithm for three-dimensional reconstruction classifies as a transform technique. It has been used for medical imaging of the human brain over the past few decades. Significant advantages of this method include: (a) its non-iterative character, (b) availability of analytical results on convergence of the solution with respect to the projection data and (c) established error estimates. The disadvantage is the large number of projections normally required for good accuracy. In engineering applications, this translates to costly experimentation, and non-viability of recording data in unsteady experiments. The use of CBP continues to be seen in steady flow experiments, particularly when the region to be mapped is physically small in size. The statement of the CBP algorithm is described as follows.

Let the path integral equation be written as

$$p(s, \theta) = \int_C f(r, \phi) dz \qquad (1)$$

where p is the projection data recorded in the experiments and f the unknown function to be determined by inverting the above equation. In practice, the function f is a field variable such as density, void fraction, attenuation coefficient, refractive index, or temperature. The symbols s, θ, r and ϕ are the ray position, view angle, position within the object to be reconstructed and the polar angle, respectively (Fig. 2). The integration is performed with respect to the independent variable z along the chord C of the ray defined by s and θ. Following [16], the *projection slice theorem* can be employed in the form

$$\bar{p}(R, \theta) = \bar{f}(R \cos(\theta), R \sin(\theta)) \qquad (2)$$

where the bar indicates the Fourier transform and R (Fourier counterpart of distance s) is the spatial frequency. The use of R follows from the literature on medical imaging. In other words,

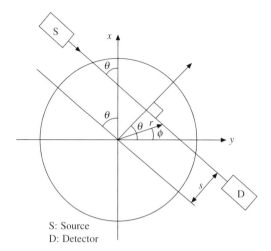

Fig. 2 **Nomenclature for the convolution back projection algorithm.**

the projection slice theorem states the equivalence of the one-dimensional Fourier transform of $p(s, \theta)$ with respect to s and the two-dimensional Fourier transform of $f(r, \phi)$ with respect to r and ϕ. A two-dimensional Fourier transform of this theorem leads to the well-known inversion formula

$$f(r, \phi) = \int_0^\pi \int_{-\infty}^{+\infty} \bar{p}(R, \theta) \exp(i2\pi Rr \cos(\theta - \phi)) |R| \, dR d\theta \qquad (3)$$

where

$$\bar{p}(R, \theta) = \int_{-\infty}^{+\infty} p(s, \theta) \exp(-i2\pi Rs) ds \qquad (4)$$

The first integral in the form given is divergent with respect to the spatial frequency R. Practical implementation of the formula replaces $|R|$ by $W(R)|R|$, where W is a window function that vanishes outside the interval $[-R_c, R_c]$. The cut-off frequency R_c can be shown to be inversely related to the ray-spacing for a consistent numerical calculation of the integral [18]. When the filter is purely of the band-pass type, Eq. (3) can be cast as a convolution integral [19]

$$f(r, \phi) = \int_0^\pi \int_{-\infty}^{+\infty} p(s, \theta) q(s' - s) ds d\theta \qquad (5)$$

where

$$q(s) = \int_{-\infty}^{+\infty} |R| W(R) \exp(i2\pi Rs) dR \qquad (6)$$

and

$$s' = r \cos(\theta - \phi) \qquad (7)$$

The inner integral over s is a one-dimensional convolution and the outer integral, an averaging operation over θ, is called *back projection*. CBP method is also known as the filtered back projection algorithm because of the filtering of the Fourier transform of the projection data $\bar{p}$ by the window (or filter) $W(R)$ (Eq. (6)). The function $q(s)$, known as the convolving function, is evaluated once and stored for repeated use for different views (or different angles). This implementation of the convolution back projection algorithm is commonly used in medical imaging.

In practical implementation of the CBP algorithm, the function $q(s)$ is determined in advance by numerical integration. Eq. (5) is subsequently evaluated (once again by numerical integration) with $p(s, \theta)$ replaced by $\tilde{T}_{ij}$, the depth-averaged temperature as a function of the ray number i and view angle j. In the present study, both i and j vary from 1 to 81. The reconstruction T_{kl} is then available for the local temperature on a $k \times l$ grid spread over a square enclosing the circular cavity.

4. Data Reduction and Uncertainty Analysis

The experiments have been performed in the infinite fringe setting of the interferometer. In the absence of any thermal disturbance, the optical path difference between the test and the reference beam is zero in the infinite fringe setting. Hence, interference is constructive and a bright field-of-view is obtained. The image obtained is practically fringe-free. When a thermal disturbance is introduced in the path of the test beam, fringes appear in the field-of-view. The fringes represents contours on which the path integral of the temperature field along the direction of the light beam is a constant. The alignment of the interferometer in the infinite fringe setting was individually carried out for the fluid phases to capture the interferograms. For this purpose, the reference chamber (containing reference test cell) was also filled with fluid layers of equal thickness as in the test cell. The reference chamber was, however, thermally inactive. The temperature drop per fringe shift ΔT_ε can be calculated from first principles of wave optics as [20]

$$\Delta T_\varepsilon = \left(\frac{\lambda / L}{dn / dT} \right) \qquad (8)$$

where λ is the wavelength of laser beam, dn/dT is the gradient of the refractive index with temperature and L the length of the test cell in the direction of the light beam. For the test cell of the present study, its length for all view angles (0, 45, 90 and 135°) remained equal. The temperature drop per fringe shift can be calculated from Eq. (8) as 5.65, 0.056 and 0.012 K, respectively, in air, water and silicone oil. Eq. (8) is applicable only when refraction errors are sensibly small. For experiments with large refraction errors, the interferograms have been qualitatively interpreted.

The fringe patterns recorded using the interferometer need to be converted first into temperature records. Quantitative evaluation of the interferograms requires that the fringe skeleton be determined. From the thinned fringes, temperature information can be generated since a fringe is an isotherm and the same temperature prevails over its entire length. In most experiments, these steps are difficult to achieve, unless suitable image processing techniques are employed [21-22].

Once the thinned fringes are obtained, calculations for the temperature distribution, interface temperature and Nusselt number (Nu) are performed. The instrumentation, image processing and data analysis employed in the present work are similar to those reported in [12] for a single fluid.

With the absolute fringe temperatures obtained, this data must be transferred to a two-dimensional uniform grid over the fluid region. This is required to apply tomographic algorithms for the reconstruction of the three-dimensional temperature field. Two-dimensional linear interpolation has been carried over the entire thinned image by superimposing a two-dimensional grid on it. The grid has 32 points along the horizontal and 21 points along the vertical direction. Once the interpolation was completed, isotherms were drawn to represent the fringes in the original image. This was done to cross-check the error involved in interpolation. It was found that the temperature data on the grid follows closely the pattern of the original thinned image and interpolation errors were negligible.

Errors in the experimental data arise from the misalignment of the apparatus with respect to the light beam, image processing operations including filtering and thinning, assigning temperatures to fringes, and interpolation and extrapolation for reconstruction. In view of a large value of ΔT_g, the number of fringes in air was small. This difficulty was circumvented in oil and water. Errors related to refraction effects in water and silicone oil were found to be high for larger cavity temperature differences, and posed a limit on the range of Rayleigh numbers (Ra) that could be studied. All experiments were conducted several times to establish the repeatability of the fringe patterns. In the event of mild unsteadiness, the dominant pattern that prevailed for the longest duration was recorded. For single fluid experiments, the plate-averaged Nusselt number has been found to be in good agreement with published correlations of [23]. Here, the correlations were coupled to an energy balance calculation that required the energy transferred across any horizontal plane to be a constant at steady state. Differences are to be expected with reference to superposed fluid layers since one of the boundaries in this configuration is a fluid interface. In a single fluid layer the top and bottom boundaries are rigid no-slip walls, while in the two-fluid configuration, one of the boundaries is a fluid interface.

The single fluid correlations employed in the present work are summarized as follows [23]:

$$\text{Nu(air)} = 1 + 1.44\left[1 - \frac{1708}{\text{Ra}}\right] + \left[\left(\frac{\text{Ra}}{5830}\right)^{1/3} - 1\right], \quad \text{Ra} < 10^6 \tag{9}$$

For water, the correlation is given by

$$\text{Nu(air)} = 1 + 1.44\left[1 - \frac{1708}{\text{Ra}}\right] + \left[\left(\frac{\text{Ra}}{5830}\right)^{1/3} - 1\right] + \frac{\text{Ra}^{1/3}}{140}\left[1 - \ln\frac{\text{Ra}^{1/3}}{140}\right], \quad \text{Ra} < 10^6 \tag{10}$$

For silicone oil (50 cSt), the correlation of large Prandtl number fluids yields

$$\text{Nu (silicone oil)} = 0.089\,\text{Ra}^{1/3} \quad (\text{Pr} \gg 1,\, \text{Ra} < 10^9) \tag{11}$$

The above equations summarize a large body of experimental data, and do not explicitly contain the aspect ratio as a parameter. Thus, the level of uncertainty associated with them is large, and specified to be ± 20%.

When applied to superposed fluid layers, Rayleigh numbers in Eqs. (9), (10) and (11) are based on the temperature difference between the interface and the nearest solid boundary. Though the interface is not an isotherm, the temperature variation over it was found to be smaller than 0.02 K, and hence negligible. Eqs. (9) to (11) can also be used to estimate the interface temperature by requiring that the energy transferred across one fluid layer is equal to that across the second, at steady state.

In air-water experiments, the Nusselt number matched the correlations of Eq. (10) in water to within ± 7% and ± 12%, respectively, at the lowest and highest Rayleigh numbers. The interface temperature matched the correlations within ± 1.8% for all the Rayleigh numbers considered in the experiments. Nusselt numbers obtained at the hot wall for each of the view angles and Rayleigh numbers were within an error band of ± 2% (Table 2). In air-oil experiments, Nusselt

Table 2. Comparison of the interface temperature (T_I) and Nusselt number (Nu) with [23] in a cavity containing layers of air and water of equal thickness (view angle marked 'V')

T, K	T_I (Exp), °C	T_I (Ref), °C	Nu		
			V = 0°	V = 90°	Ref
6.5	28.00	28.86	3.95	3.96	3.50
8.5	28.66	29.72	4.55	4.65	3.74

number at the hot wall in oil and interface temperatures matched the correlations to within ± 5% and ± 0.5%, respectively, for the lower range of Rayleigh numbers (Table 3). At the highest Rayleigh number, the uncertainty in measurements became quite high due to a very large number of fringes and high refraction errors.

Table 3. Comparison of the interface temperature (T_I) and Nusselt number (Nu) with [23] in a cavity containing layers of air and silicone oil of equal thickness (view angle marked 'V')

T, K	T_I (Exp), °C	T_I (Ref), °C	Nu		
			V = 0°	V = 90°	Ref
0.4	29.93	30.03	1.17	1.18	1.12
1.8	29.80	29.87	1.83	1.88	1.70

5. Results and Discussion

Convection in horizontal superposed layers differs from that in a single layer mainly due to the formation of a well-defined interface. If one of the fluids is a gas, the interface could be a free surface, and the motion of the fluids could be practically uncoupled. When the fluids are placed in an enclosure and the motion is thermally driven, flow patterns in the fluid layers are coupled even when one of the phases is a gas. In a two-layer system, instabilities and convection patterns depend on several dimensionless groups. Specifically fluid property ratios, Rayleigh numbers of the individual fluid layers, layer thickness ratio, cavity aspect ratio and the overall geometry of the confining surfaces are important.

The planform of the cellular pattern is determined by the shape of the apparatus. In the circular cavity, the rolls are expected to form concentric rings [24]. The fringes (namely, the contours of constant depth-averaged temperature) obtained arrange themselves to form a symmetric Ω (omega) pattern. The full thermal field of the fluid layer is a collection of omegas and inverse omegas inter-linked with one another. Such patterns have also been noticed in experiments and calculations for buoyancy-driven convection in a large aspect ratio circular cavity filled with a single fluid [25].

The appearance of Ω-shaped fringes has been interpreted as being equivalent to an axisymmetric temperature field. The interferograms have been subsequently interpreted to understand the influence of increasing Rayleigh number on transition to three-dimensionality and unsteadiness. In addition, average interface temperature and cavity Nusselt number have been determined and are compared with the correlations [23] in Tables 2 and 3. The thermal fields discussed in the following pertain to the convection patterns that form after the passage of the initial transients.

5.1 Three-dimensional Structures

Temperature fields on selected planes of the fluid layers have been reconstructed from its inteferometric projections obtained at view angles of 0, 45, 90 and 135°. The projection data obtained at a view angle of 180° (fifth view) is identical to that of the 0° view angle and has also been used. As the cavity is nominally circular, the principles of radial tomography has been applied. Convolution back projection (CBP) has been chosen as the tomographic algorithm in the present work and is discussed in Section 3.1.

A large number of projections are required to apply effectively the reconstruction algorithm. In the present experimental set-up, five view angles alone could be recorded in the form of interferometric projection data. Further, projection data obtained from a view angle covered 41% of the full width of the fluid layer in the cavity. The partial data has been obtained in the central portion of the cavity in the experiments. As required in tomography, the laser beam should scan the full width of the thermal field in each view. To achieve the requirement, the projection data was suitably extrapolated by a numerical scheme. The data size thus generated was a square matrix of size 81(rays) × 81(view angles). The extrapolated data covered the full width of the cavity. The grid size is a reasonably large data set in terms of discretization. The extrapolation errors are not expected to be serious since the interferometric measurement is less sensitive towards the edges of the test cell where the geometric path length is small. Information that is de-emphasized in one projection is, however, captured in the next. In this respect, the thermal field is fully represented by the partial data itself. The process of extrapolation casts it in a form that is suitable for the numerical implementation of CBP.

The original projection data is available at five view angles, resulting in an angular increment of 45°. It is known that a coarse increment, common to *limited-data* tomography reconstructs the major features of the thermal field, while the minor features (in terms of physical size) are lost [26]. For this reason, tomography has been used in the present study to examine axisymmetry of temperature distribution, but not the small scale features such as knots and spokes.

For applying tomographic algorithm, the interferograms for all view angles, namely 0, 45, 90 and 135° have been evaluated quantitatively through image analysis techniques to determine the temperature field in the fluid layers [27]. The reconstructed temperature field in the individual fluid layers using CBP algorithm are now presented for selected experiments.

5.1.1 Air-water Experiments

Steady state interferograms for all view angles are shown in Figs. 3 (a-d) and 4 (a-d), respectively, for temperature differences of 6.5 K and 8.5 K across the cavity. The Rayleigh numbers are 9,530 (air), 48,474 (water) at 6.5 K and 12,466 (air), 66,992 (water) at 8.5 K. The fields shown correspond to the size of a single optical window. The window dimension in the experiments was 41% of the nominal cavity diameter. View angles of 0, 45, 90 and 135° are considered. The interferograms were recorded after an experimental run time of 4 hours, when the thermal field was fully evolved and the fringe patterns were quite steady. The top and bottom surfaces have been maintained at the temperatures of 22.5 and 29°C, respectively, leading to a temperature difference of 6.5 K across the cavity. While, for the temperature difference of 8.5 K, the bounding surfaces were at 21.5 and 30°C. Based on the temperature field evaluated quantitatively through image analysis techniques, the estimated experimental interface temperature and the average Nusselt number have been calculated and are compared with the single-fluid correlations in Table 2 for ΔT = 6.5 and 8.5 K.

Three horizontal planes, namely y/h = 0.15, 0.5 and 0.85 have been considered for reconstruction,

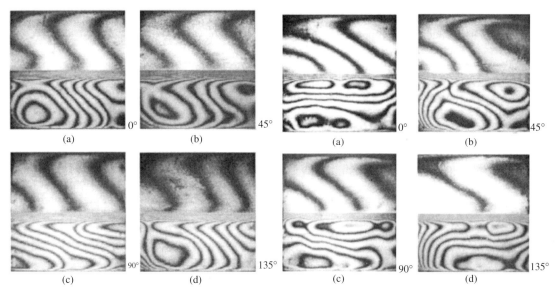

Fig. 3 Long-time interferograms obtained in a cavity filled with air and water for a temperature difference of 6.5 K. View angles: (a) 0°, (b) 45°, (c) 90° and (d) 135°.

Fig. 4 Long-time interferograms obtained in a cavity filled with air and water for a temperature difference of 8.5 K. View angles: (a) 0°, (b) 45°, (c) 90° and (d) 135°.

where y is the vertical coordinate and h the water layer height. The projection temperatures obtained at the three planes have been plotted with respect to the transverse coordinate for all view angles for a temperature difference of 6.5 K and 8.5 K. Here, the transverse coordinate is parallel to the width of the cavity, and $y/h = 0$ represents the hot wall of the cavity.

At steady state, the average temperature over a given plane should be a constant for all the view angles. The projection data obtained along the transverse coordinate show inconsistency in this respect though the differences are within the limits of experimental errors.

The constancy of the average temperature has been strictly enforced by linear scaling of the projection data before applying the tomographic algorithm. For the correction, the grand average of all the projection data over a given plane was calculated. The average of projection data for each view angle was then forced to be equal to the grand average. With this approach, the individual trends for each view angle were retained and at the same time, the average temperature was a constant for a given plane, with respect to the individual projections. This consistent data set has been supplied as an input for the reconstruction of the three-dimensional thermal field.

The reconstruction of the temperature field in water over three horizontal planes from its interferometric projection is shown in Fig. 5 (a-f). For the lower Rayleigh number in Fig. 5 (a, c, e), a degree of axisymmetry is to be noticed in the form of concentric rings in the central portion of the cavity. The concentricity of the thermal field in the central region is seen to decrease from the near wall ($y/h = 0.15$) to the interface region ($y/h = 0.85$). The temperature contours show the center to be at a lower temperature as compared to its surrounding region. There is also a progressive reduction in the size of the isotherm of $T = 27.4\,°C$. This indicates that the cooler fluid descends at the center while the warmer fluid ascends at a neighbouring location to form a full roll (plume model). At a higher Rayleigh number (Fig. 5 (b, d, f)) the extent of axisymmetry is uniform. This is surprising because convection at $\Delta T = 8.5$ K was found to be strongly time-dependent. The similarity of the

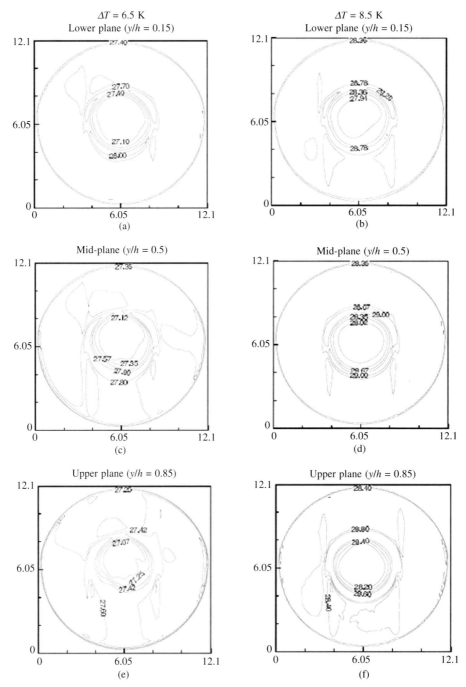

Fig. 5 Reconstructed temperature contours in the water layer. Cavity is half-filled with water, the rest being air. Temperature differences are 6.5 K (a, c, e) and 8.5 K (b, d, f) respectively, at lower-, mid- and upper-plane of the cavity.

reconstructed temperatures shows that the dominant interferograms recorded for each view angle correspond to a particular roll-structure, that in turn is close to axisymmetric. The sign of circulation within the roll is indeterminate because no broadening of the isotherm (for example, $T = 28.4°C$) is to be seen, going from one plane to the next.

5.1.2 Air-silicone Oil Experiments

Experiments in an octagonal cavity containing superposed layers of air and silicone oil are discussed in this section. For measurements, temperature differences of 0.4 and 1.8 K were applied across the cavity. At higher temperature differences, the unresolved high density fringes and the refraction in the oil layer restricted the analysis of interferograms. Fringes for view angles of 0, 45, 90 and 135° were recorded in the experiments for reconstructing the three-dimensional thermal field in the oil layer where sufficiently large number of fringes were obtained.

Fig. 6 (a-d) shows the interferograms recorded for a temperature difference of 0.4 K across the cavity. The two bounding cold and hot surfaces were maintained at temperatures of 29.7 and 30.1°C, respectively. The corresponding Rayleigh numbers in air and oil were 470 and 2011, respectively. The temperature drop per fringe shift in air being 5.65 K, the above cavity temperature differences did not produce any fringe in the air portion of the cavity.

Fringe patterns in the oil phase (Fig. 6 (a-d)) show an Ω-pattern. This indicates the thermal field to be axisymmetric about the vertical axis passing through the center of cavity. Thus, the flow developed in the oil layer classifies as steady two-dimensional in the regime map of [28]. The fringe density is quite high near the lower wall, indicating that the rolls are closer to the interface. In air, no fringes were obtained, though a conduction state is to be expected owing to the Rayleigh number being less than the critical value of 1708.

In Fig. 7 (a-d), the temperature difference of 1.8 K was imposed across the cavity containing

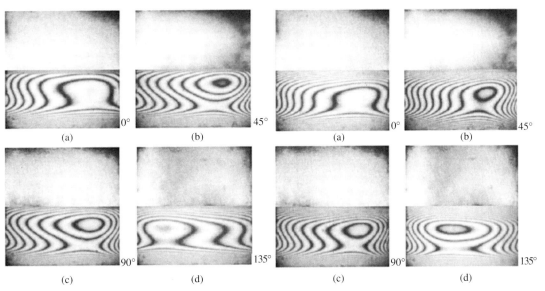

Fig. 6 Interferograms in a cavity filled with air and silicone oil. Cavity temperature difference is 0.4 K. View angles: (a) 0°, (b) 45°, (c) 90° and (d) 135°.

Fig. 7 Interferograms in a cavity filled with air and silicone oil. Cavity temperature difference is 1.8 K. View angles: (a) 0°, (b) 45°, (c) 90° and (d) 135°.

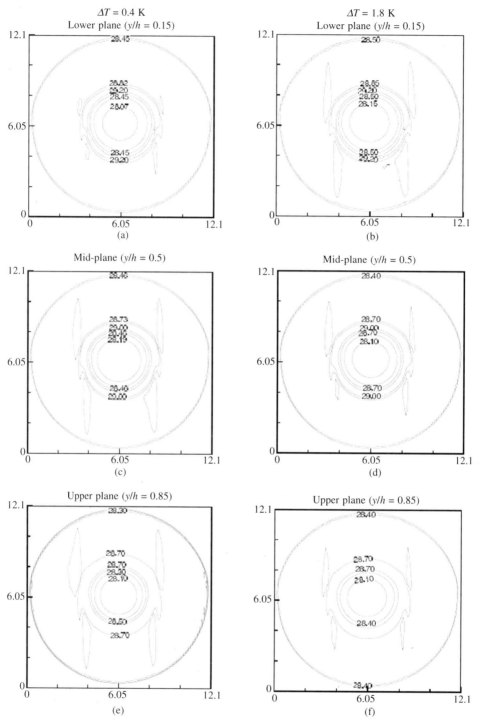

Fig. 8 Reconstructed temperature contours in the oil layer in the cavity half-filled with silicone oil, the rest being air at temperature differences of 0.4 K (a, c, e) and 1.8 K (b, d, f), respectively, at lower-, mid- and upper-plane of the cavity.

equal layer heights of air and oil, respectively, for the view angles of 0, 45, 90 and 135°. The experimental and the estimated interface temperatures calculated were 29.80 and 29.87°C, respectively. Accordingly, the Rayleigh numbers were determined to be 2237 and 6963, respectively, in the air and the oil layers. An increase in the Rayleigh numbers led to vigorous convection in the cavity. The thermal boundary layer formed near the lower wall was of lower thickness, indicating an increase in the roll size. The number of fringes in oil was higher due to an overall increase in the temperature drop between the lower wall and the interface. Once again fringes were not seen in air. Steady Ω-shaped isotherms, though with a distortion were obtained in the oil layer (Fig. 7 (a-d)).

Reconstructed temperature contours from its interferometric projection in oil are shown in Fig. 8 (a-f) for the temperature differences of 0.4 and 1.8 K for the three horizontal planes selected. In Fig. 8, the reconstructed field for the temperature differences of 0.4 K (a, c, e) and 1.8 K (b, d, f) have been shown for the lower ($y/h = 0.15$), mid ($y/h = 0.5$) and upper ($y/h = 0.85$) planes. Near-concentric rings are obtained in the central portion of the cavity for all the planes. This is a confirmation of axisymmetry in the projection data obtained from the experiments. Concentricity is higher at the mid- and the upper-plane as compared to the lower plane. This is owing to the presence of rolls in those planes, while a thermal boundary layer is formed near the lower plane. The temperature contour closer to the center is seen to be at a lower temperature compared to its immediate neighbour. For cavity temperature differences of 0.4 and 1.8 K, the central isotherm obtained at the respective plane is seen to decrease in size from the lower- to upper-plane. The isotherms at the lower- and the mid-planes are almost equal in size. The plume model proposed for water is applicable in the present context as well.

6. Conclusions

Buoyancy-driven convection in two superposed fluid layers have been studied using laser interferometry. Tomography was applied to interferograms recorded with a Mach-Zehnder interferometer. The analysis in the individual fluid layers was carried out in the post-critical range of Rayleigh numbers. The following conclusions have been arrived at in the present work.

Based on the projection data obtained in the cavity containing air and water, the fluid layers showed a degree of axisymmetry in the thermal field at a lower Rayleigh number. With an increase in Rayleigh number, three-dimensionality was seen in water while axisymmetry persisted in air. In the air-oil experiments, omega-shaped fringe patterns indicating a steady two-dimensional thermal field was obtained in the oil layer over the range of Rayleigh numbers studied.

Three-dimensional reconstruction at three selected planes in the fluid layers confirmed the thermal field to be axisymmetric wherever the circular isotherms on a plane was obtained. In the air-oil experiments, near-concentric rings are obtained in terms of the temperature contours in the central portion of the cavity. Deviation from concentricity is seen at higher cavity temperature differences. Thus, the thermal field in silicone oil becomes increasingly non-axisymmetric with increasing Rayleigh number. The breakdown of axisymmetry is, however, delayed in the near-wall region.

References

1. Rasenat, S., Busse, F.H. and Rehberg, I., A theoretical and experimental study of double-layer convection, *J. Fluid Mechanics*, **199**, pp. 519–540, 1989.

2. Dijkstra, H.A., On the structure of cellular solutions in Rayleigh-Benard-Marangoni flows in small-aspect-ratio containers, *J. Fluid Mechanics*, **243**, pp. 73–102, 1992.
3. Busse, F.H. and Sommermann, G., Double-layer convection: a brief review and some recent experimental results. In: *Advances in Multi-fluid Flows*, Editors: Y.Y. Renardy, A.V. Coward, D.T. Papageorgiou and S.-M. Sun, SIAM Publication, pp. 33–41, 1996.
4. Andereck, D.C., Colovas, M.M. and Peter, W.D., Observation of time-dependent behaviour in the two-layer Rayleigh-Benard system, *Proceedings of the Third Microgravity Fluid Physics Conference*, pp. 313–318, 1996.
5. Golovin, A.A., Nepomnyashchy, A.A. and Pismen, L.M., Nonlinear evolution and secondary instabilities of Marangoni convection in a liquid-gas system with deformable interface, *J. Fluid Mechanics*, **341**, pp. 317-341, 1997.
6. Prakash, A. and Koster, J.N., Steady Rayleigh-Benard convection in a two-layer system of immiscible liquids, *ASME J. Heat Transfer*, **118**, pp. 366–373, 1996.
7. Prakash, A., Yasuda, K., Otsubo, F., Kuwahara, K. and Doi, T., Flow coupling mechanism in two-layer Rayleigh-Benard convection, *Experiments in Fluids*, **23**, pp. 252–261, 1997.
8. Johnson, D. and Narayanan, R., Experimental observation of dynamic mode switching in interfacial-tension-driven convection near a codimension-two point, *Physical Review E*, **54**, No. 4, pp. 3102–3104, 1996.
9. Johnson, D., Narayanan, R. and Dauby, P.C., The effect of air height on the pattern formation in liquid-air bilayer convection, *Fluid Dynamics at Interfaces*, Cambridge University Press, Cambridge, pp. 15-30, 1999.
10. Mayinger, F. (Editor), *Optical Measurements: Techniques and Applications*, Springer-Verlag, Berlin, 1994.
11. Michael Y.C. and Yang K.T., Three-dimensional Mach-Zehnder interferometric tomography of the Rayleigh-Benard problem, *ASME J. Heat Transfer*, **114**, pp. 622–629, 1992.
12. Mishra, D., Muralidhar, K. and Munshi, P., Interferometric study of Rayleigh-Benard convection at intermediate Rayleigh numbers, *Fluid Dynamics Research*, **25**, No. 5, pp 231–255, 1999.
13. Mishra, D., Muralidhar, K. and Munshi, P., Interferometric study of Rayleigh-Benard convection using tomography with limited projection data, *Experimental Heat Transfer*, **12**, No. 2, pp. 117–136, 1999.
14. Mishra, D., Muralidhar, K. and Munshi, P., Performance evaluation of fringe thinning algorithms for interferometric tomography, *Optics and Lasers in Engineering*, **30**, pp. 229–249, 1999.
15. Kirchartz, K.R. and Oertel H. (Jr), Three-dimensional thermal cellular convection in rectangular boxes, *Journal of Fluid Mechanics*, **192**, pp. 249–286, 1988.
16. Herman, G.T., *Image reconstruction from projections*, Academic Press, New York, 1980.
17. Muralidhar, K., Temperature field measurement in buoyancy-driven flows using interferometric tomography, *Annual Review of Heat Transfer*, **12**, Edited by Chang-Lin Tien, Vishwanath Prasad and Frank Incropera, pp. 265-375, 2002.
18. Munshi, P., Application of computerized tomography for measurements in heat and mass transfer, *Proceedings of the 3rd ISHMT-ASME Heat and Mass Transfer Conference*, IIT Kanpur (India), 29-31 December 1997, Narosa Publishers, New Delhi.
19. Ramachandran, G.N. and Laxminarayanan, A.V., Three-dimensional reconstruction from radiographs and electron micrographs: Application of convolution instead of Fourier transforms, *Proc. Nat. Acad. USA*, **68**, pp. 2236–2240, 1970.
20. Goldstein, R.J. (Editor), *Fluid mechanics measurements*, Hemisphere Publishing Corporation, New York, 1983 (second edition: 1996).
21. Gonzalez, R.C. and Woods, R.E., *Digital image processing*, Addison-Wesley Publishing Company, USA, 1993.
22. Jain, A.K., *Fundamentals of digital image processing*, Prentice-Hall International Editions, USA, 1989.
23. Gebhart, B., Jaluria, Y., Mahajan, R.L. and Sammakia, B., *Buoyancy-induced flows and transport*, Hemisphere Publishing Corporation, New York, 1988.

24. Velarde, M.G. and Normand, C., Convection, *Scientific American*, **243**, No. 1, pp. 79–94, 1980.
25. Srivastava, A. and Panigrahi, P.K., A combined numerical-experimental study of convection in an axisymmetric differentially heated fluid layer, *Indian Journal of Engineering and Materials Sciences*, **9**, pp. 448–454, 2002.
26. Natterer, F., *The mathematics of computerized tomography*, John Wiley & Sons, New York, 1986.
27. Punjabi, S., Muralidhar, K. and Panigrahi, P.K., Buoyancy-driven convection in two superposed fluid layers in an octagonal cavity, *International Journal of Thermal Sciences*, **43/9**, pp. 849–864, 2004.
28. Krishnamurti, R., Some further studies on the transition to turbulent convection, *J. Fluid Mechanics*, **60**, pp. 285–303, 1973.

Computerized Tomography for Scientists and Engineers
Edited by P. Munshi
Anamaya Publishers, New Delhi, India

10. Tomography in Fusion Plasma Research

C.V.S. Rao
X-ray Diagnostics Group, Institute for Plasma Research,
Bhat, Gandhinagar-382 428, India

Abstract

Tomography has demonstrated its usefulness to diagnose fusion plasmas. Its non-invasive character has made it attractive in plasma physics. It is essentially a technique to recover the three-dimensional distribution from the measured line integral projections through a mathematical integral transform like Abel transform. Many plasma diagnostics are made capable of measuring line integrals of plasma parameters for tomographic purpose. Some commonly used techniques are microwave interferometer, Faraday rotation, electron cyclotron emission, visible, ultraviolet, and x-ray emission etc. These diagnostics are being used to study plasma properties like equilibrium, stability, plasma-wall interaction and impurity content, MHD phenomenology, particle and energy transport and turbulence in the plasma core. In fact, tomography has revolutionized the field of plasma diagnostics and has lots of promise for the future fusion experiments. This article presents tomography in fusion experiments highlighting some of the salient features.

---

1. Introduction

The first published work on tomography was as early as 1826 by N.H. Abel, a Norwegian physicist, for an object with axi-symmetrical geometry. In 1917 Austrian mathematician Johann Randon extended Abel's idea for objects with arbitrary shapes [1]. Allen Cormack, Tuft University, analyzed the problem in terms of an expansion in circular harmonics in 1964 [2]. In 1965, R.N. Bracewell working in the field of Radio Astronomy showed that line integrals are related to Fourier transform of the source function and by Fourier techniques deduced the inversion [3]. A commercial laboratory EMI Ltd. in Great Britain (1971) announced the development of the EMI Scanner—a symbiotic marriage of x-ray scanning system and digital computer technology. Godfrey Hounsfield generated images of isolated slices of the brain with exquisite discrimination in 1973. Myers and Levine for the first time in fusion plasma diagnostics performed 2-D spectral line emission reconstruction in 1978 [4]. In 1979, Godfrey Hounsfield and Allen Cormack were jointly awarded the Nobel Prize for their pioneering work on x-ray tomography. Sauthoff and Von Goeler (1979) demonstrated for the first time [5] in fusion plasma research that x-ray tomography gives useful information about the magnetic topology. They reconstructed the x-ray emissivity contours using series expansion method.

2. Fusion Plasma Concepts

The most promising fusion reaction is the one using heavier isotopes of hydrogen. Deuterium and tritium nuclei fuse together to produce alpha particle with a release of a neutron and generation of 17.6 MeV of energy (Fig. 1)

$$^2_1D + ^3_1T \Rightarrow ^4_2He\ (3.5\ \text{MeV}) + ^0_1n\ (14.4\ \text{MeV})$$

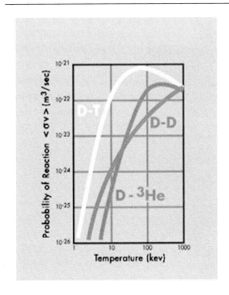 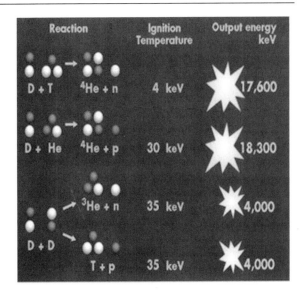

Fig. 1

In order to induce fusion of nuclei of deuterium and tritium it is necessary to overcome the mutual repulsion due to their positive charges.

The promising method of supplying the energy is to heat the deuterium and tritium fuel to sufficiently high temperature so that the thermal velocities of the nuclei are high enough to produce the required reactions. Fusion brought about this way is called *thermonuclear fusion*. At such temperatures the fuel is fully ionized. The electrostatic charge of the ion is neutralized by the presence of an equal number of electrons and the resulting neutral gas is called *plasma*. At such temperatures no material wall can confine plasma particles. Tokamak fusion device provides a method of confining plasma particles by magnetic field. Tokamak is a Russian invention by two Nobel Laureates 'Andrei Sakharov' and 'Igor Tamm'. Tokamak is simply an abbreviation for the more precise specification of this invention 'toroidal chamber with magnetic coils' and is the simplest configuration for plasma magnetic confinement (Fig. 2).

Field holds particles in small gyrating orbits. It is possible by this means to arrange ions to travel large distances—'million times the dimensions of the vessel before reaching the wall'.

Lawson's criteria: From the balance of power input to produce the thermonuclear fusion and the power losing from the reaction, we get criteria $n\tau_E$ (product of ion density and the energy confinement time), which must be around $n\tau_E \approx 1.5\text{-}3.0 \times 10^{20}$ m^{-3}s and temperature of the ion at $T_i = 10\text{-}20$ keV.

The required product $n\tau_E T_i$ is 3.0×10^{21}m^{-3}s keV.

Components of the magnetic fields in a Tokamak, which confines particles and energy, are as follows (Fig. 3):

(i) Toroidal magnetic field produced by coils placed all around the torus.
(ii) Poloidal magnetic field produced running a current in the plasma.
(iii) Vertical magnetic field produced by coils placed outside to keep the plasma column in equilibrium.

The net result of the two components (i) and (ii) of the magnetic fields is helical field lines.

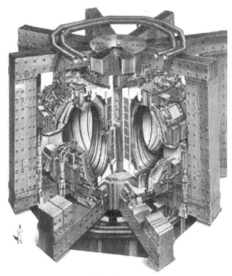

Fig. 2

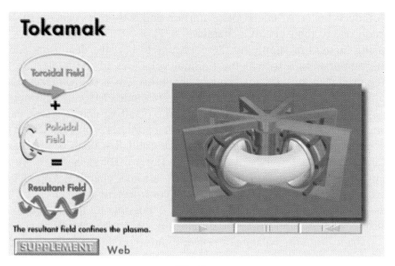

Fig. 3

The twist introduced in the field lines takes the particles from inside to outside and vice-versa, thus shorting the electric field produced due to charge separation, which in-turn is created because of drift of particles in pure Toroidal field. Component (iii) of the Tokamak magnetic field, viz. the vertical field, is applied to arrest the hoop force produced by current carrying channel in the plasma.

2.1 Magnetic Surfaces

Particle starting from some point on a poloidal plane returns to that plane after traversing round the torus and misses the point from where it started off by an infinitesimally small amount. After the particle executes many revolutions round the torus it forms surfaces. These surfaces, formed by the magnetic field lines, are the very basis of magnetic confinement schemes. These nested surfaces thus formed are responsible for confining plasma particles and energy inside the Tokamak. In addition to these surfaces there are locations where the magnetic field lines go once round the torus and returns to the same point on the poloidal plane from where it started. In contrast to the nested surfaces these are called *rational surfaces*. In fact these are not surfaces at all. Rational surfaces are formed when the field lines go around the torus once to complete one revolution in the poloidal plane ($q = m/n = 1/1$). Then there are field lines which go round twice the long way to complete once in the small way ($q = m/n = 2/1$) and similarly three times the long way to go round once the small way ($q = m/n = 3/1$), and so on (Fig. 4).

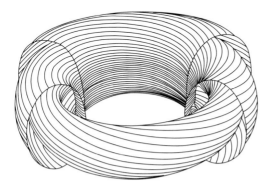
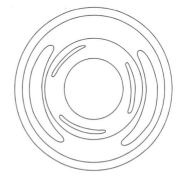

Fig. 4

Plasma particles are confined by the formed nested surfaces and at the rational surfaces it is most vulnerable to slightest MHD perturbation. Tokamak plasma is a state of matter, which is far from thermodynamic equilibrium. It has many free sources of energy in the form of gradients of pressure and plasma current density. These sources feed energy into instabilities, which then grow. They are small and local structures, which quickly grow into larger structures and become global in nature and are mainly responsible for the loss of particles and energy from the plasma in these devices.

These effects manifest themselves when the nice nested surfaces are destroyed by the so-called 'tearing modes' tearing or snapping of field lines and later reconnection of the same forms the basis of magnetic island formation, i.e. the main plasma channel carrying current gets broken-up into number of filaments due to this type of instability. These smaller channels of current produce magnetic fields, which interact with that produced by the main channel resulting in magnetic islands (Fig. 5).

The main role of tomography in plasma diagnostics is to clearly bring out the detailed history of these features—the fast growing structures, their spatial and temporal evolutions.

Tomography is essentially search for a solution of an inverse problem, where a given property of the system (such as radiative emission, blood flow or fall in temperature) has to be reconstructed starting from a set of line integrated measurements. We can view the problem as a transformation of the local emission in (r, θ)-space into a line-integrated brightness in (p, ϕ)-space or reconstruction of the source function from the projections (Fig. 6).

Tomographic techniques have been carried out with various probes such as x-rays, γ-rays, visible light, microwaves, electrons, protons, neutrons, heavy ions, ultrasound and nuclear magnetic

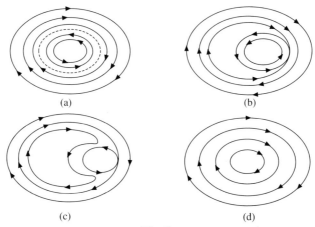

Fig. 5

resonance. They have been used to study a large variety of objects whose sizes vary over an enormous range, from complex molecules studied by molecular biologists to distant radio/pulsar sources studied by radio astronomers. Among the vast number of objects that fall between these extremes, human body or some particular organ in the body is an especially important case. It has been used in various disciplines such as medicine, astronomy, molecular biology, geophysics, seismology, electron microscopy, material science, holography, aerodynamics, fluid dynamics, nuclear reactor physics and fusion plasma research.

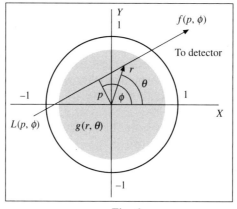

Fig. 6

3. Difference Between Medical and Plasma Tomography

Much of the foundation of the principle of tomographic imaging has been pioneered by research in the field of medicine from which fusion diagnostics work borrows heavily. However, there are some important differences between the two applications discussed as follows.

The fusion diagnostic measurement is on x-ray emission whereas in medicine it is x-ray absorption. Therefore, instrumentation is significantly different, though, in principle the mathematics of the problem remains same. The reconstruction is in 2-D planar slice corresponding to a cross section at one Toroidal location. This slice has a finite width in the Toroidal direction, defined by the collimating optics, and Toroidal variation of the emissivity over this width will be unresolved and could create errors in the reconstructed image. Unlike the 'pencil-beam' absorption chords referred to in medical x-ray imaging, the collimation necessary for emission tomography means that the chord lines are actually solid angles. The finite beam width effects and size of the solid angle sets one of the lower limits on the spatial resolution along with chord spacing. The density of measurements is one of the major differences between medical imaging and Tokamak imaging. In modern CT-scanners, used for medical diagnosis, x-ray absorption measurements are made on about 300,000 different chords arranged in a very regular pattern, whereas in Tokamak applications the largest system operating so far has 310 chords arranged in a typically irregular pattern. This huge disparity is partly due to the fact that external access to a patient, unlike Tokamak plasma, is unhindered from any angle, and partly due to the lack of necessity for good time resolution in medical imaging. This disparity is so large that the software algorithms, which work best for medical tomography, are not useful for imaging Tokamak plasmas. Finite element or filtered back projection method is best suited for medical imaging. Analytical solutions (series expansion methods) with appropriate constraints (discretize before it is inverted) built-in the algorithm are suitable for plasma imaging. Additional information like irregular coverage due to restricted access and need for additional information to supplement the limited number of measurement (symmetries in the problem etc.) can be easily incorporated.

4. Tomography Systems at Institute for Plasma Research (IPR)

We have developed a number of systems that are already operational and deployed on the Aditya Tokamak while some are in process of development. The soft x-ray (SXR) tomography system is already operational. Bolometer and microwave interferometer have recently started operating. The VUV-imaging, Optical tomography and Hard x-ray tomography systems are in the developmental stage. The description on the soft x-ray tomography system is discussed as follows.

5. Soft X-ray Tomography System

5.1 Principle

Tokamak plasma emits radiations due to electron-ion collisions and is dominated by Bremsstrahlung radiation. The spectral region in which the greatest Bremsstrahlung power is emitted is in the vicinity of photon energy equal to the electron temperature. For $h\nu \sim T_e$ radiation in this optimum spectral range for hot plasma falls in the soft x-ray region (1-20 keV) is sufficient and convenient for rapid time-resolved observations of its evolution to be made. The emissivity $g(r, \theta)$ depends upon various crucial plasma parameters such as electron temperature, electron density and impurity content. Emissivity is a constant on magnetic surfaces. The aim is to obtain relative variations of this complicated combination of plasma parameters that combine to determine the emissivity as a function of time and space. This enables us to see qualitatively, and on occasions quantitatively, the evolution of the shape of the plasma as a result of instabilities or other perturbations. Therefore, contours of emissivity should coincide with magnetic surfaces and a plot as the one obtained from soft x-ray tomography can be considered as approximately a magnetic surface plot (Fig. 7).

The basic principle of radiative emission tomography is that of measuring the emitted radiation from a plasma poloidal cross section along a large number of collimated lines-of-sight. The plasma is optically thin for these wavelengths. The lines are defined by arrays of detectors and pinholes and are identified by their impact parameter p (perpendicular distance from the origin to the line-of-sight) and chord angle ϕ (the angle that the normal to the line-of-sight makes with the equatorial plane).

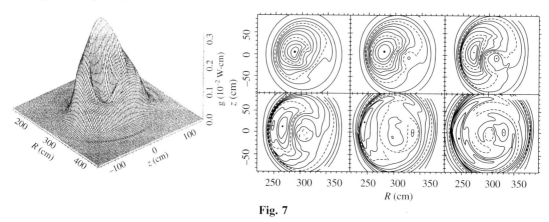

Fig. 7

The access to the plasma allowed by the aperture in the plasma vacuum vessel, which are usually rather tiny and often to be shared with other diagnostics systems, sets a limit to the number of lines-of-sight. With several measurements of brightness $f(p, \phi)$, defined as the line integral of the emissivity $g(r, \theta)$ along a specific line-of-sight $L(p, \phi)$

$$f(p, \phi) = \int_{L(p,\phi)} g(r, \theta) dl$$

Dynamic changes of the field line topology should be visible on SXR tomograms as changes in the topology of emissivity contours. In general, the instrument intrinsically limits the radial and poloidal resolution of the tomographic results. The number and orientation of different views (arrays) determine the poloidal resolution and the number of chords per view and their overlap determines the spatial resolution.

5.2 Instrumentation

5.2.1 System Hardware

Aditya Tokamak is equipped with two array cameras in the pinhole geometry—horizontal and vertical. The horizontal camera is mounted on a radial diagnostic port of the Aditya machine with its axis passing through the midplane of the device while vertical array is mounted on a top port at the same toroidal angle but displaced poloidal by 90° (Fig. 8).

Each array consists of 18 detectors, silicon surface barrier diodes, the radiation-exposed area of 50 mm^2 and a thickness of 100 μm (depletion layer). There is an aluminium entrance window in ohmic contact with the silicon wafer. This whole assembly is encapsulated in a gold plated enclosure. The diodes are operated in the current-mode at room temperature and negatively biased to 9 volts.

The detectors respond to radiation in the energy range 0.3 to 15 keV that can be restricted by different filters to choose proper energy spectrum and optimize the dynamic range. The radiation is collimated by circular aperture of diameter 4.0 mm which produces a fan-like viewing configuration as shown in Fig. 9.

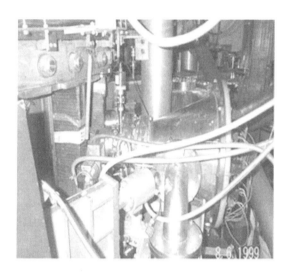

Fig. 8

The chord spacing and the size of viewing cones defined by the diodes and aperture—neighbouring viewing cones roughly touching each other—fix a lower limit for the spatial resolution of the system. It is about 1.6 cm at the vessel center.

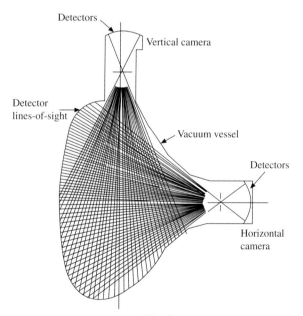

Fig. 9

Few of the detectors are purposely blinded, i.e. covered with stainless steel absorbers in order to measure the background noise. All the diodes are calibrated using x-ray source *in-situ*.

5.2.2 Signal Conditioning (Analog) Electronics and Data Acquisition System

There are severe problems in air-core machine, where remarkable stray fields and sources of electromagnetic noise are present. The simultaneous requirements of high time resolution, good spatial accuracy, high precision in the measurements, and the need for fast processing of large amount of data are met by the dedicated high performance analog electronics for signal conditioning and the general purpose data acquisition system.

5.2.3 Signal Conditioning Electronics

In each diode a current is generated in response to the photon flux incident, typically below few hundreds of nanoamperes, which is amplified by a low noise, wide-band preamplifier with a transimpedance of 10^5 volts/amp. The preamplifier stage consists of a current-to-voltage convertor, an amplifier stage and a line driver. The signals from the preamplifier are carried to the main amplifier, 50 m away placed in the diagnositics bay, by means of specially made low noise twisted pair cables with two braided shields. The amplifier and filtering stage consists of the main amplifier, a filter and opto-coupler. To reduce ground loops, the system common ground is defined at, and only at, the input of each preamplifier. The array cone is at machine potential. To reduce capacitive coupling and radiofrequency noise above 500 Hz, each array is further shielded with a 3 mm thick copper Faraday cage inside the array cone connected to the diagnostic system ground.

5.2.4 Data Acquisition System

LeCroy CAMAC modules digitize the signals, LRS-2264 (8 bits, 8 inputs waveform digitizers) at a sampling rate of 125 kHz. The signal-to-noise ratio in the central channel is better than 30. The digitized data is stored in memory modules that can be accessed by computer interfaced with it. Reconstruction of emissivity profiles using pre-processed data to be performed either on-line, within few minutes of the pulse, or off-line with higher time resolution (Fig. 10).

Platform independent codes are required for tomographic inversion. The reconstruction is

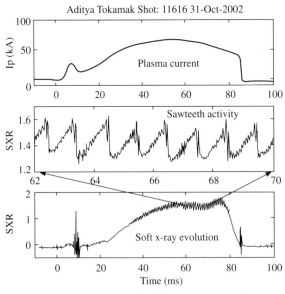

Fig. 10

performed in VME CPU's running real-time operating system on Linux workstations. For this, the whole code has been written in 'C' language, using its standard library only, so that transferring the system to new machine is straightforward.

6. Reconstruction or Inversion Method

It has been shown that the measured signal by each detector can be written as

$$f(p, \phi) = \int_{L(p,\phi)} g(r, \theta) dl \tag{1}$$

where $g(r, \theta)$ is the plasma emissivity and $f(p, \phi)$ is the brightness. The emitting region, which may have a cross section of any arbitrary shape, is assumed to be lying in a circle of radius normalized to unity. The local emission is to be reconstructed over this area, specified by $0 \leq r \leq 1$ and $0 \leq \theta \leq 2\pi$.

The 2-D quantities in Eq. (1) are decomposed into their Fourier harmonics to take care of the angular coordinates

$$g(r, \theta) = \sum_{m=0}^{\infty} [g_m^c(r) \cos(m\theta) + g_m^s(r) \sin(m\theta)] \tag{2}$$

$$f(p, \phi) = \sum_{m=0}^{\infty} [f_m^c(p) \cos(m\phi) + f_m^s(p) \sin(m\phi)] \tag{3}$$

with the radial coordinates expanded as follows.

Substituting these expansions in Eq. (2), we get

$$f_m^{c,s}(p) = \sum_{l=0}^{\infty} a_{m,l}^{c,s} f_{m,l}(p) \tag{4}$$

$$g_m^{c,s}(r) = \sum_{l=0}^{\infty} a_{m,l}^{c,s} g_{m,l}(r) \tag{5}$$

$$f_{m,l}(p) = 2 \int_p^1 \frac{g_{m,l}(r) T_m(p/r) r \, dr}{\sqrt{(r^2 - p^2)}} \tag{6}$$

where $f_{m,l}(p)$ is related to $g_{m,l}(r)$ and $T_m(x)$ is the Chebyshev polynomial of the first kind. McCormack had expanded $g(r, \theta)$ in terms of a complete set of orthogonal functions known as Zernicke polynomials which are found to generate artifacts at the edge of the plasma where the emission is low. Therefore, we have used Fourier-Bessel expansion to represent the source function (*Pramana*, **61**, 141(2003)).

The final form reduces to

$$f(p, \phi) = -\sum_{m=0}^{\infty} \sum_{l=0}^{\infty} [a_{m,l}^c \cos(m\phi) + a_{m,l}^s \sin(m\phi)]$$

$$\times \left[-2\sqrt{(1-p^2)} J_m'(x_{m,l}) \sum_{n=0, n \neq m}^{\infty} \delta_n J_n(x_{m,l}) \sin(n\pi/2 - x_{m,l} p) \times \left\{ \frac{U_{m+n-1}(p)}{(m+n)} + \frac{U_{m-n-1}(p)}{(m-n)} \right\} \right] \tag{7}$$

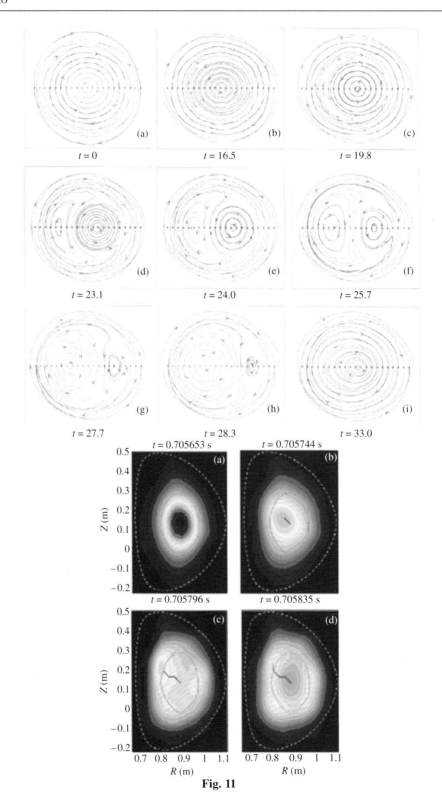

Fig. 11

The base functions for poloidal harmonics and radial expansion must be truncated at some maximum values of m and l in order to solve for unknowns.

The highest m-value that can be reconstructed is about equal to the number of detector arrays and maximum l-number which can be used depends upon the sampling density of chords within a projection or an array, i.e. chord spacing in p coordinate.

Equation (7) is a linear expansion of $f(p, \phi)$ into a finite number of terms representing 2-D surfaces, each of which has an unknown coefficient. If $f(p,\phi)$ is measured along M different chords to obtain M, then the independent linear equation can be written in matrix form as

$$aW = f$$

The elements of W matrix depend only on detector geometry and the number of angular and radial harmonics, which are used to fit the data. If these do not change, then whole matrix can be calculated once. As long as the number of M detectors exceed the number of unknown coefficients, we can arrive at a solution for these unknowns in the least square sense by a simple and efficient matrix multiplication

$$a = W^{-1} f$$

These coefficients are then used in Eq. (4) to get $g_m^{C,S}(r)$. These Fourier emission harmonics are then summed up in Eq. (2) to give the total reconstructed image (Fig. 11).

7. Conclusions

Sawtooth activity and internal disruptions characterized by the mode structures $m = 0$, $m = 1$ and $m = 2$ close to disruptions have been routinely monitored in Aditya Tokamak. Coupling of $m = 1$ and $m = 2$ islands as precursor to major disruptions is studied in-depth.

Fast real time data acquisition has been achieved. Time resolution of the order of tens of microseconds is obtained on routine basis. At present the spatial resolution is of the order of 16 mm and at the plasma centre it is being upgraded to 8 mm. We are able to handle low level signals of the order of few nanoamperes in the hostile Tokamak environment.

There are extremely critical needs of the fusion community, which in future have to be addressed, i.e. to understand the physical mechanism of the micro instabilities to detect them and to control them in real time. To be able to control the position and shape of the plasma cross-section in real time. There is a strong felt need for networking between various groups to make significant progress in this area.

References

1. J. Randon, *ber Ges Wissenschaften Leipzig*, **69**, p. 262 (1917).
2. A.M. Cormack, *J. Appl. Phys.*, **34**, p. 2722 (1963).
3. R.N. Bracewell and A.C. Riddle, *Astrophys. J.*, **150**, p. 427 (1967).
4. B.R. Myers and M.A. Levine, *Rev. Sci. Instru.*, **49**, p. 610 (1978).
5. N.R. Sauthoff and S. von Goeler, *IEEE Transactions on Plasma Science*, **PS-7**, p. 141 (1979).

Computerized Tomography for Scientists and Engineers
Edited by P. Munshi
Anamaya Publishers, New Delhi, India

11. Tomographic Reconstructive Techniques for Void Fraction Distribution in Heavy Density Liquid Metal Two-phase Flows

P. Satyamurthy, N.S. Dixit, R. Chaudhary and P. Munshi*

Laser & Plasma Technology Division, Bhabha Atomic Research Centre, Mumbai-400 085, India
*Department of Mechanical Engineering, Indian Institute of Technology, Kanpur-208 016, India

Abstract

Heavy density molten liquid metals lead, lead-bismuth-eutectic (LBE) and mercury have been proposed as spallation targets for neutron source with gas induced circulation of these targets. To design these systems, it is required to understand two-phase high-density flows under various flow regimes and develop suitable CFD codes, which are to be validated with experimental data. The two-phase flow regimes (bubbly, Churn, Slug) are governed by void fraction distribution and the accurate determination of void fraction in these systems is complicated in view of high-density, opacity. A 6.5 m high mercury facility has been set up in our institute to study two-phase nitrogen and mercury flows at ambient temperature. The flow is upward and co-current. The flow regime obtained in the facility (bubbly and Churn) is similar to the flows expected in the gas-driven spallation targets. A non-intrusive 7-beam (Fan beam) and a parallel beam void fraction measurement systems based on high-energy gamma-rays (^{60}Co of 50 and 100 mCi source) for these flows have been developed. The attenuation of these gamma-rays, which are passed across the cross section of the flow area are measured under three conditions, viz. when the pipe is filled with mercury, when filled with nitrogen and when there is a two-phase flow present. The data is collected at different orientations by rotating and parallel displacement of the source and detectors. Studies have been carried out with many variants of Algebraic Reconstruction Technique (ART) and Multiplicative Algebraic Reconstruction Techniques (MART) algorithms for reconstruction of void fraction distribution. All these algorithms were tested by comparing with predicted values of simulated void fraction distribution. In general, MART algorithms were better suited than ART for void fraction reconstruction for heavy density two-phase liquid metal flows.

1. Introduction

High density liquid metal two-phase flows are finding applications in many advanced technologies like gas driven spallation target advanced reactors, power conversion from low-grade heat sources by MHD etc. [1]. For neutron spallation source, some of the candidate liquid metals are lead, lead-bismuth-eutectic and mercury. To design these systems, it is required to understand two-phase high-density flows in the riser of the loop under various regimes and develop suitable CFD codes, which are to be validated with experimental data. The two-phase flow regimes (bubbly, churn, slug) are governed by void fraction distribution and the accurate measurement of void fraction in these systems is complicated in view of high density and opacity.

Void fraction in liquid metal flows can be measured by intrusive probes based on electromagnetic (potential, micro-magnetic) and opto-mechanical principles [4, 5] or by non-intrusive methods using gamma-rays for high density [6] and X-rays for low-density systems [4]. Non-intrusive probes have the advantage of being used under all process conditions (high temperature, corrosive fluid etc.) and do not interfere with the flow system. In this article, details of high energy gamma-ray attenuation system, developed to measure two-dimensional void distribution in the nitrogen-mercury flow based on: (i) single-detector parallel beam and (ii) seven-detector fan-beam system are presented. Various tomographic methods developed for the reconstruction of void profiles are also presented. Most suitable reconstruction methods suitable for determining void distribution in heavy density liquid metal two-phase flow are suggested.

2. Mercury-Nitrogen Experimental Facility

A 6.5 m high mercury facility has been set up in Bhabha Atomic Research Centre (BARC), Mumbai to study two-phase nitrogen and mercury flows at ambient temperature. The flow is upward and co-current. The flow circulation is achieved by creating two-phase flow in the riser. The flow regime obtained in the facility consists of bubbly and Churn flow. The loop consists of mixer, riser pipe (internal diameter 79.0 mm), separator, down-comer etc. Nitrogen is introduced through the mixer at ~6 kg/cm^2 pressure. A two-phase mercury-nitrogen mixture is established in the riser, which gives rise to density difference between the riser and down-comer leading to the circulation of liquid metal in the loop. Nitrogen is separated and is let out to the ambient. Mercury alone flows through the down-comer. A maximum of 60 kg/s mercury flow rate has been achieved. Void fraction is measured using fan beam gamma-ray source at 1.2 m from the mixer and with single beam parallel chords at 2.9 m from the mixer (Fig. 1).

Fig. 1 Mercury-nitrogen liquid metal MHD experimental facility at BARC.

3. Single Parallel Beam High Energy Gamma-ray System

This system, located at 2.9 m above the mixer is shown in Fig. 2. The gamma-ray source having 3 mm window for the beam along with the detector system was installed on a horizontally movable and rotating platform. ^{60}Co of activity 50 mCi was used as gamma-ray source and 1.33 MeV photons were chosen for measurements. NaI (Th) detector (located at a distance 225 mm from the source) with PMT, pre-amplifier etc. are mounted along with source to collect the projection data from single gamma-ray beam. The measurements were taken at 14 chord lengths (parallel to one another by moving source and detector) at a given orientation when the pipe was filed with mercury, when empty and during two-phase flow. This was repeated for every 30° of rotation covering 180° (6-views). One set of experiment generated 84 line averaged void fraction values for the given flow conditions.

Fig. 2 Parallel and rotational single beam scanning gamma-ray system (^{60}Co of 50 mCi).

4. Fan Beam High Energy Gamma-ray System

This system, shown in Fig. 3, is installed at a distance of 1.2 m from the mixer. ^{60}Co of 100 mCi gamma-ray source is housed in an appropriate lead container. The radioactive source container

Fig. 3 Multi-detector high energy gamma-ray system for the measurement of void distribution.

consists of three lead blocks with one blind, the second one with 7 windows each of 3 mm diameter and the third housing the source, which can be moved in front of the window to obtain fan beam. NaI (Th) detectors (located at a distance 725 mm from the source) with PMT, pre-amplifier etc. are mounted along with source to receive fan beam gamma-rays (the angle between two adjacent rays is 3.5° and equally spaced from the centre line). These rays pass through various chords of the flow in a circular pipe on a rotating platform. Gamma-ray attenuation was measured along the 7 chord lengths for every 10° interval up to 170°, when the pipe was filled with mercury, when empty and during two-phase flow. Based upon this data, void distribution was determined with appropriate reconstruction algorithms. 1.33 MeV beam of photons was chosen for these measurements. One set of experiment generated 126 line averaged void fraction values for given set of flow conditions.

5. Experiment

Experiments were carried out for many flow rates. In this article data corresponding to 4.7 g/s flow rate of nitrogen and 43.0 kg/s flow rate of mercury is presented. Photon counts for each measurement were taken in excess of 3000 to reduce the Poisson corruption [7]. The size of the gamma-ray beam was chosen to be 3 mm so that the error due to finite beam size was negligible [8].

6. Determination of Line Averaged Void Fraction

From the measured data, the line averaged void fraction, $\overline{\beta}_i$ for each ray was obtained from the following relation.

$$\overline{\beta}_i = \ln(I_i^t/I_i^l)/\ln(I_i^g/I_i^l)$$

where $\overline{\beta}_i$ is the ith chord average void fraction, I the number of photon counts per unit time, t, l, g correspond to two-phase flow, mercury alone and nitrogen alone, respectively, and i is the ith ray.

7. Tomographic Inversion Techniques

Two-dimensional field is divided into square cells, and cells are numbered in a regular fashion from bottom to top. The void fraction is assumed to be constant within the cell (i.e. α_j, is the void fraction of jth cell). In the present analysis, source, detector, and rays are considered ideal. The length of intersection of ith ray and j^{th} cell, denoted by $w_{i,j}$ for $i = 1, 2 \ldots M$ and $j = 1, 2 \ldots N$ is the path length of ith gamma-ray in the jth cell. Schematic of the discretization of the physical domain is shown in Fig. 4.

By accounting the contribution of each cell into the average void fraction along a ray, the result is the system of linear equations

$$\overline{\beta}_i = \frac{\sum_{j=1}^{N} w_{i,j}\, \alpha_j}{\sum_{k=1}^{N} w_{i,k}}$$

$$\overline{\alpha}_i = \overline{\beta}_i \sum_{k=1}^{N} w_{i,k}$$

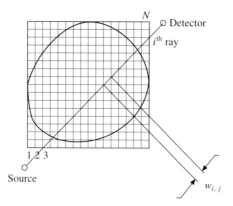

Fig. 4 Discretization of field.

If M and N were small, conventional matrix theory methods can be used to invert the system of equations. However, in practice M and N are large enough which precludes any possibility of direct matrix inversion. Of course, when noise is present in the projection data, and when $M < N$, even for small N it is quite difficult to use direct inversion method; some least square methods may have to be used. But when M and N are large, such methods are also impractical. Some iterative series expansion and optimization methods can be used for the inversion of the above problem.

7.1 Series Expansion and Optimization Inversion Techniques

The incomplete projection data leads to the application of series expansion and optimization methods for the present work. The statements for algorithms with additive corrections are given first, followed by multiplicative correction algorithms and in the end optimization methods have been presented.

7.1.1 Additive ART

Simple ART
Simple ART step by step is as follows [9]:
    Initialization $f^0 \in R^n$ is arbitrary.
    do for each iteration k
    do for each angle of projection θ
    do for each ray i_θ
    (a) Compute the correction $((\overline{\alpha}_{i_\theta})_{\exp} - (\overline{\alpha}_{i_\theta})_{\text{theo}})$

    (b) Compute the total value of weight function along each ray, $W_{i_\theta} = \sum_{j=1}^{N} w_{i_\theta,j}$

    (c) Calculate the average weight function $\dfrac{(\overline{\alpha}_{i_\theta})_{\exp} - (\overline{\alpha}_{i_\theta})_{\text{theo}}}{W_{i_\theta}}$

    end do
    do for each ray i_θ
    (a) Update each field parameter as

$$f_j^{new} = f_j^{old} + \lambda \frac{w_{i_\theta,j}((\overline{\alpha}_{i_\theta})_{exp} - (\overline{\alpha}_{i_\theta})_{theo})}{W_{i_\theta}}$$

end do
do for each ray i_θ
(a) Update approximate projections using

$$(\overline{\alpha}_{i_\theta})_{theo} = \sum_{j=1}^{N} w_{i_\theta,j} f_j, \quad i_\theta = 1, \ldots M_\theta$$

end do
end do

(b) Stop when abs$\left(\frac{(f_j^{k+1} - f_j^k)}{f_j^k}\right) \times 100 \leq 0.01$

end do

Gordon ART

In this method corrections are applied to all the cells through which i^{th} ray passes, before calculating the correction for the next ray. Hence, the number of rays per angle of irradiation is not important [10].

Initialization $f^0 \in R^n$ is arbitrary.
do for each iteration k
do for each ray i
(a) Compute the correction $((\overline{\alpha}_i)_{exp} - (\overline{\alpha}_i)_{theo})$
(b) Compute the correction coefficient

$$c_i = \sum_{j=1}^{N} w_{i,j}^2$$

(c) Apply a correction to each cell j of the test field through which the present ray passes as

$$f_j^{new} = f_j^{old} + \lambda \frac{w_{i,j}((\overline{\alpha}_i)_{exp} - (\overline{\alpha}_i)_{theo})}{c_i}$$

end do

(a) Stop when abs$\left(\frac{(f_j^{k+1} - f_j^k)}{f_j^k}\right) \times 100 \leq 0.01$

(b) Compute projection numerically

$$(\overline{\alpha}_i)_{theo} = \sum_{j=1}^{N} w_{i,j} f_j, \quad i = 1, \ldots M$$

end do

Gilbert ART

This method is also known as SIRT (simultaneous iterative reconstruction technique). In SIRT, the elements of the field function are modified after all the correction values corresponding to individual rays have been calculated [11].

Initialization $f^0 \in R^n$ is arbitrary.
do for each iteration k
do for each ray i
(a) Compute the correction $((\overline{\alpha}_i)_{exp} - (\overline{\alpha}_i)_{theo})$
(b) Compute the correction coefficient

$$c_i = \sum_{j=1}^{N} w_{i,j}^2$$

end do
do for each ray i
(a) Identify all the rays, M_{cj} passing through a given cell and the corresponding $w_{i,j}$ and $((\overline{\alpha}_i)_{exp} - (\overline{\alpha}_i)_{theo})$.
(b) Apply a correction to each cell j of the test field through which the present ray passes as

$$f_j^{new} = f_j^{old} + \sum_{i=1}^{M_{cj}} \lambda \frac{w_{i,j}((\overline{\alpha}_i)_{exp} - (\overline{\alpha}_i)_{theo})}{c_i}$$

end do

(a) Stop when abs $\left(\frac{(f_j^{k+1} - f_j^k)}{f_j^k} \right) \times 100 \leq 0.01$

(b) Compute projection numerically

$$(\overline{\alpha}_i)_{theo} = \sum_{j=1}^{N} w_{i,j} f_j, \quad i = 1, \ldots M$$

end do

Anderson ART
This algorithm combines the ART and SIRT algorithms. Method of applying correction is similar to SIRT but the structure is similar to ART [12].
Initialization $f^0 \in R^n$ is arbitrary.
do for each iteration k
do for each angle of projection θ
do for each ray i_θ
(a) Compute the correction $((\alpha_{i_\theta})_{exp} - (\overline{\alpha}_{i_\theta})_{theo})$
(b) Compute the total value of weight function along each ray

$$c_{i_\theta} = \sum_{j=1}^{N} w_{i_\theta,j}^2$$

end do
do for each ray i_θ
(a) Update each field parameter as

$$f_j^{new} = f_j^{old} + \lambda \frac{w_{i_\theta,j}((\alpha_{i_\theta})_{exp} - (\overline{\alpha}_{i_\theta})_{theo})}{c_{i_\theta}}$$

end do
do for each ray i_θ

(a) Update the approximate projections using

$$(\bar{\alpha}_{i_\theta})_{\text{theo}} = \sum_{j=1}^{N} w_{i_\theta, j} f_j, \quad i_\theta = 1, \ldots, M_\theta$$

end do
end do

(a) Stop when abc $\left(\dfrac{(f_j^{k+1} - f_j^k)}{f_j^k} \right) \times 100 \leq 0.01$

end do

Multiplicative ART

When the correction is multiplicative, the ART is called multiplicative ART (MART) [10, 13, 14].

Initialization $f^0 \in R^n$ is arbitrary.
do for each iteration k
do for each ray i
(a) Identify all the rays, M_{cj} passing through a given cell and the corresponding $w_{i,j}$, $(\bar{\alpha}_i)_{\text{exp}}$ and $(\bar{\alpha}_i)_{\text{theo}}$
(b) Apply a correction to each cell j of the test field through which the present ray passes as

GBH MART $\quad f_j^{\text{new}} = f_j^{\text{old}} \times \prod_{i=1}^{M_{cj}} \left[1 - \lambda \left(1 - \dfrac{(\bar{\alpha}_i)_{\text{exp}}}{(\bar{\alpha}_i)_{\text{theo}}} \right) \right]$

GH MART $\quad f_j^{\text{new}} = f_j^{\text{old}} \times \prod_{i=1}^{M_{cj}} \left[1 - \lambda \dfrac{w_{i,j}}{w_{\max}} \left(1 - \dfrac{(\bar{\alpha}_i)_{\text{exp}}}{(\bar{\alpha}_i)_{\text{theo}}} \right) \right]$

Lent MART $\quad f_j^{\text{new}} = f_j^{\text{old}} \times \prod_{i=1}^{M_{cj}} \left[\dfrac{(\alpha_i)_{\text{exp}}}{(\bar{\alpha}_i)_{\text{theo}}} \right]^{\frac{\lambda w_{i,j}}{w_{\max}}}$

Lent2 MART $\quad f_j^{\text{new}} = f_j^{\text{old}} \times \prod_{i=1}^{M_{cj}} \left[\dfrac{(\alpha_i)_{\text{exp}}}{(\bar{\alpha}_i)_{\text{theo}}} \right]^{\lambda w_{i,j}}$

end do

(a) Stop when abs $\left(\dfrac{(f_j^{k+1} - f_j^k)}{f_j^k} \right) \times 100 \leq 0.01$

(b) Compute projection numerically

$$\bar{\alpha}_i = \sum_{j=1}^{N} w_{i,j} f_j \quad i = 1, \ldots, M$$

end do

7.2 Optimization Techniques

7.2.1 Maximum Entropy Method

The probability of finding a system in a given state depends upon the multiplicity of that state. That is to say, it is proportional to the number of ways you can produce that state. Here a 'state' is defined by some measurable property, which would allow you to distinguish it from other states. A greater value of multiplicity of state implies greater possibility of the system in that state. Based on the above-mentioned definition of entropy a tomographic algorithm is developed by Gull [15]. This method produces an unbiased solution and is maximally non-committal about unmeasured parameters. A formulation for the present algorithm is given as follows:

Maximum entropy functional
$$F = -\sum_{j=1}^{N} f_j \ln(f_j) \tag{3.1}$$

subject to the constraints
$$\overline{\alpha}_i = \sum_{j=1}^{N} w_{i,j} f_j \quad i = 1, \ldots, M \text{ (projection data)} \tag{3.2}$$

and $f_j \geq 0$ (a prior condition).

The Lagrangian multiplier technique has been used for maximization of the functional F under imposed constraints. The problem now reduces to the solution of a set of non-linear equations with unknown Lagrangian multipliers. Non-linear equations are solved after linearization using Taylor series expansion method by Gauss-Seidel iterative technique [16].

7.2.2 Minimum Energy Method

When a system has minimum energy, then system is in most stable state. Based on the above definition formulations for minimum energy method is given as follows [15]:

Minimum energy functional is defined as
$$F = -\sum_{j=1}^{N} f_j^2 \tag{3.3}$$

subject to the constraints

$$\overline{\alpha}_i = \sum_{j=1}^{N} w_{i,j} f_j \quad i = 1, \ldots, M \quad \text{(projection data)} \tag{3.4}$$

The Lagrangian multiplier technique has been used for minimization of the functional F under imposed constraints. The problem is now reduced to the solution of a set of linear equations with Lagrangian multipliers unknown. Standard Gauss-Seidel method is used to solve the linear equations in Lagrangian multipliers and then field values have been calculated [16].

8. Simulated Results

The performance of the aforesaid discussed algorithms has been checked over simulated data. Mainly three field functions have been used to test performance constant, step, and impulse inside a unit circle. 61 rays and 12 views in 180° total view angle have been used to generate the projection data. Ray spacing during reconstructions has been taken as 0.0164. During all reconstructions the initial value in each pixel has been taken 1.

Original fields (constant, step, impulse) are given in Figs. 5 (a), 6 (a) and 7 (a) along with their reconstructions using different tomographic algorithms in Figs. 5 (b-j), 6 (b-j) and 7 (b-j).

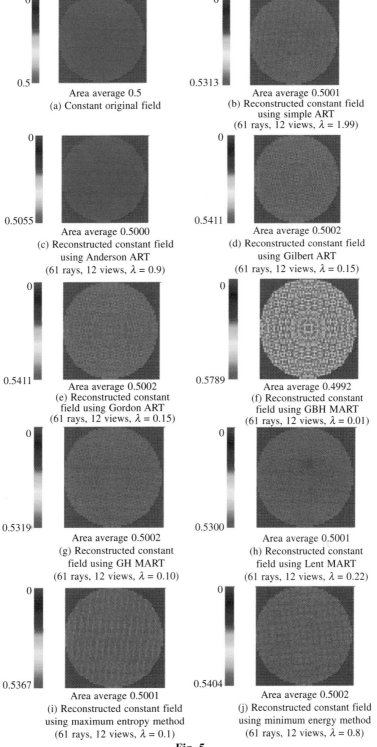

Fig. 5

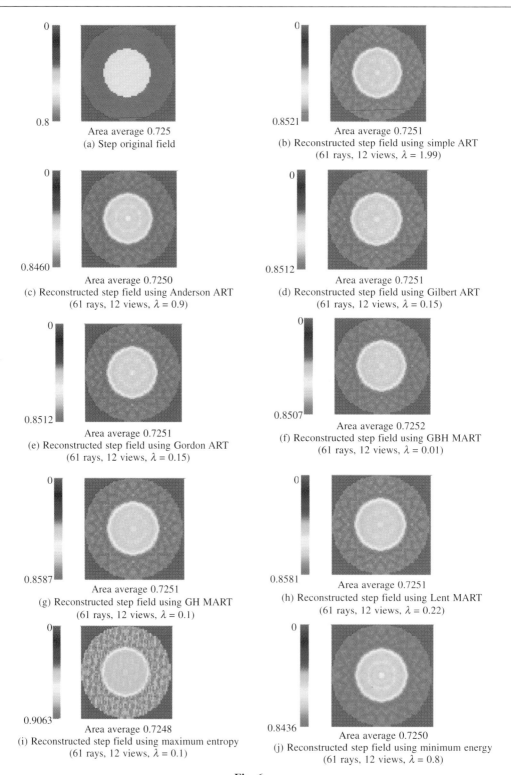

Fig. 6

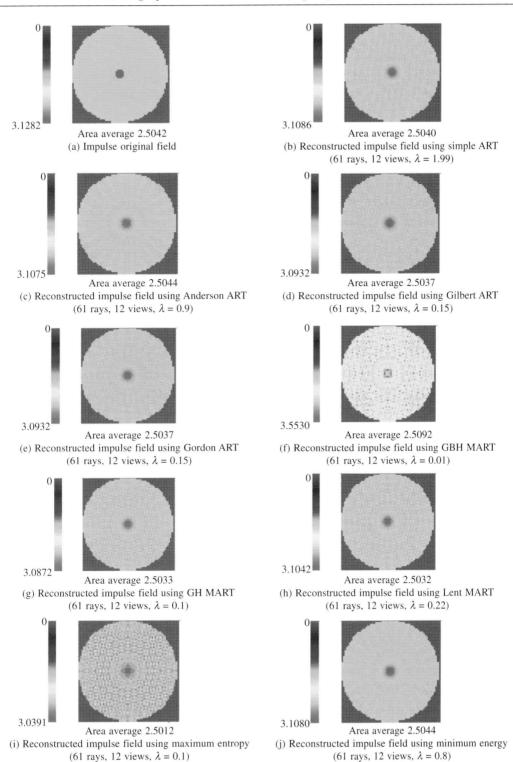

Fig. 7

9. Determination of Void Fraction Distribution in Nitrogen-Mercury Flow

Void fraction distributions at 2.9 m and 1.2 m from the mixer have been calculated using different MART algorithms. The void fraction at upper level is given in Fig. 8 (a, b) with the GH and GBM MART, respectively. The maximum value of the void fraction was obtained at the central region with a value of 0.479 and the lower value near the wall region at the higher level. The average value across the cross-section comes out to be 0.35. As expected the flow is essentially approaching the fully developed flow [17].

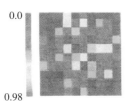

0.98
Average void fraction = 0.352
(a) Void fraction distribution by GH MART
parallel-beam at 2.9 m from the mixer
(Flow rate: nitrogen = 4.7 g/s and mercury = 43 kg/s)

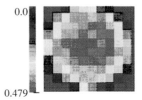

0.479
Average void fraction = 0.349
(b) Void fraction distribution by GBH MART
parallel-beam at 2.9 m from the mixer
(Flow rate: nitrogen = 4.7 g/s and mercury = 43 kg/s)

Fig. 8

The measured void distribution in fan beam mode for the same flow rates at the location 1.2 m are given in Fig. 9 (a, b) which shows an average value nearly 0.18. Fig. 9 (a, b) also shows that the overall void fraction is much lower as compared to upper location for the same flow rate. This is due to higher pressure at this location leading to higher gas density. Further, we see that highest void fraction obtained was nearly 0.624 in one cell somewhere not in central region. This value may be spurious due to limitation of number of rays. This has to be further analyzed.

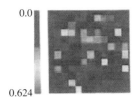

0.624
Average void fraction = 0.181
(a) Void fraction distribution by GH MART
fan-beam at 1.2 m from the mixer
(Flow rate: nitrogen = 4.7 g/s and mercury = 43 kg/s)

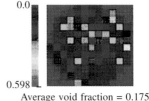

0.598
Average void fraction = 0.175
(b) Void fraction distribution by GBH MART
fan-beam at 1.2 m from the mixer
(Flow rate: nitrogen = 4.7 g/s and mercury = 43 kg/s)

Fig. 9

10. Conclusions and Future Scope

Non-intrusive void fraction measurement based on gamma-ray attenuation method is a powerful diagnostic technique for two-phase heavy density liquid metal flows. In this article, details of void fraction measurements based on high-energy gamma ray attenuation method along with various reconstruction tomographic techniques used for two-phase heavy density liquid metal flows are presented. These flows give rise to unique features with respect to measurement as well as reconstructive algorithms. This is primarily because of very high attenuation even for high energy gamma-rays (>1 MeV). Because of this there are certain limitations in the data collection: large collection time for each measurements adds up large background counts leading to errors in the counts as well as possibilities of drift in the electronic circuits (drift in the high voltage power supply can lead to shift in the gamma-ray peak etc). Large collection time also puts limitations on the number of angles and rays for data collection and practical limitation for application to transient flows (requirement of kilo-Curie sources for collecting minimum number of counts (~3000) in millisecond time scale for a typical flow). In view of this limitation of the data, not every reconstructive algorithm is suitable for these systems. Our analysis has found that most suitable algorithms are MART and currently we are looking into the applicability of Maximum Entropy and Minimum Energy algorithms.

Development of these tomographic techniques will lead to accurate determination of void profiles in complex systems like LMMHD, ADS reactor, Spallation Target loop and other systems containing heavy density two-phase flows. This information is vital for optimum design of these systems.

References

1. Bauer, G. S., "Physics and technology of spallation neutron sources", Nuclear Instruments and Methods in Physics Research, A 463 (2001) 505–543.
2. Energy Amplifier Demonstration Facility Reference Configuration, Summary Report, Ansaldo, EAB0.00 1200-Rev.0, Italy, Jan. 1999.
3. Satyamurthy, P. and Biswas, K., "Design of a LBE spallation target for fast-thermal accelerator driven sub-critical system (ADS)", 7th Information Exchange Meeting, Jeju, Korea, 14-16, October (2002).
4. Eckert, S., Gerbeth, G., Guttek, B., Strechemesser, H. and Lielausis, O., "Investigations of liquid metal two-phase flow characteristics by means of local resistivity probes and X-ray screening techniques", International Workshop on Measuring Techniques for Liquid Metal Flows, Rossendorf, Germany, Oct. 11-13 (1999).
5. Cartellier, A., "Optical probes for local void fraction measurements: Characterization of performance", *Rev. Sci. Instruments*, **61(2)**, (1990), 874–886.
6. Thiyagarajan, T.K., Satyamurthy, P., Dixit, N.S., Venkatramani, N., Garg, A. and Kanvinde, N.R., "Void fraction profile measurements in two-phase mercury-nitrogen flows using gamma ray attenuation method", *Experimental Thermal and Fluid Science*, **10** (1995), 347–354.
7. Munshi, P. and Vaidya, M.S., "A Sensitivity Study of Poisson Corruption in Tomographic Measurements in Air Water Flows", *Trans. Am. Nucl. Soc.* **68**, June (1993), 234–235.
8. Satyamurthy, P., Thiyagarajan, T.K., Dixit, N.S. and Venkatramani, N., "Void fraction measurement in two-phase liquid metal flows: Correction for finite beam size of the gamma rays", Second Int. Conf. on Energy Transfer in Magneto Hydro Dynamic Flows, Aussois, France (1994), 597–606.
9. Mayinger, F., "Optical Measurements", Springer-Verlag, New York, 1994.
10. Gordon, R., Bender, R. and Herman, G.T., "Algebraic reconstruction technique (ART) for three-dimensional electron microscopy and X-ray photography", *J. Theo. Biol.* **29** (1970), pp. 471–481.
11. Gilbert, P.F.C., "Iterative methods for three-dimensional reconstruction of an object from its projections", *J. Theo. Biol.* **36** (1972) pp. 105–117
12. Anderson, A.H. and Kak, A.C., "Simultaneous algebraic reconstruction technique (SART): a superior implementation of the ART algorithm", *Ultrason. Imaging* **6** (1984).

13. Verhoeven, D., "Multiplicative algebraic computed tomographic algorithms for the reconstruction of multidimensional interferometeric data", *Opt. Eng.* **32** (1993) pp. 410–419.
14. Gordon, R. and Herman G.T., "Three-dimensional reconstruction from projections: A review of algorithms", *Int. Rev. Cytol*, **38** (1974), pp. 111–115.
15. Gull, S.F., "Maximum entropy tomography", *Appl. Opt.* **25** (1986), pp. 156–160.
16. Mishra, D., Murlidhar, K. and Munshi, P., "A Robust Mart Algorithm for Tomographic Applications", Numerical Heat Transfer, Part B 35, 1999, 485–506.
17. Neal, L.G. and Bankoff, S.G., "Local parameters in co-current mercury-nitrogen flows", *AIChE Journal*, **11** (1965), 624–635.
18. Chaudhary, R. "Tomographic Techniques and Simulation of High-density Two-phase Flows for Gas-Driven ADS Target Systems", M.Tech Thesis, IIT Kanpur, India, Aug. 2003.

Computerized Tomography for Scientists and Engineers
Edited by P. Munshi
Anamaya Publishers, New Delhi, India

12. Imaging of Buoyancy-Driven Convective Field Around a KDP Crystal Using Schlieren Tomography

Atul Srivastava, K. Muralidhar and P.K. Panigrahi
Department of Mechanical Engineering, Indian Institute of Technology Kanpur,
Kanpur-208 016, India

Abstract

Buoyancy-driven convection around a KDP crystal growing from its aqueous solution has been imaged by a laser schlieren technique. Buoyancy-driven flow in the solution is initiated by inserting a KDP seed into its supersaturated solution that is followed by slow cooling of the solution. The deposition of the excess salt from the solution changes the concentration field around the crystal and initiates convection in the growth cell. The experiments have been conducted in a circular cavity of diameter 19 cm and height 30 cm. Monochrome schlieren, an optical technique has been employed to scan the flow field. The convection pattern has been recorded from four different view angles (0°, 45°, 90° and 135°) by turning the crystal growth chamber. Quantities of interest in the measurement are concentration distribution in the vicinity of the growing crystal and concentration profiles. The optical technique employed shows quite clearly the spatial distribution of concentration differences in the vicinity of the growing crystal. With the passage of time, the solution in the growth cell was found to develop stable density stratification, leading to a practically stagnant solution. By interpreting the schlieren images as projection data of the solute concentration, the three-dimensional concentration field around the growing crystal has been determined using tomographic algorithms. Convolution back projection (CBP) and combined ART-CBP tomography algorithms have been employed for the reconstruction of the concentration field at different horizontal planes in the growth chamber. The reconstructed concentration fields reveal the symmetry of flow field around the growing crystal for the range of parameters studied.

1. Introduction

Optical measurement techniques employed in fluid flow, heat and mass transfer applications can be classified into three main categories: (1) direct visualization, where a type of marker (e.g., a dye) follows the fluid particles; (2) detection of frequency shift of the scattered illumination from seeded particles, and (3) measurement of the distribution of index of refraction of the medium under study, from which properties such as density, temperature and concentration can be determined. Optical techniques are non-intrusive in nature and are practically inertia-free. A majority of them are field techniques in the sense that an entire cross-section of the physical region can be mapped. They require the physical medium to be transparent and are thus suitable for measurement in fluids.

Optical methods that utilize the dependence of refractive index of light on the quantity to be measured such as temperature and concentration can be mainly divided into three categories:

1. *Interferometry*, where the fringe formation is related to the changes in the refractive index with respect to a reference environment.
2. *Schlieren*, where light deflection in a variable refractive index field is captured.
3. *Shadowgraph*, where the reduction in light intensity on beam divergence is employed.

For fluids, a unique relationship can be established between refractive index and the local density. If the bulk pressure of the region is practically constant, density in turn correlates with temperature and concentration, and the three methods become applicable for the measurement of these quantities. Optical techniques yield the path integral of the local density of the medium (or its spatial derivatives) under study. The three-dimensional temperature and concentration fields can be reconstructed from the projection data using the principles of tomography.

The sensitivities of the three optical techniques referred above are quite different. Specifically, interferometry is suitable for applications where the density gradients are small. Schlieren and shadowgraph find applications where large density gradients are present in the flow field. Interferometric images provide data on the temperature and concentration distributions. In contrast, schlieren and shadowgraph respond, respectively, to the first and second derivatives of the density field. Consequently, they have found utility in the past, principally as flow visualization tools. It is now possible to precisely quantify changes in light intensity, they can be extended to provide quantitative measurements as well.

A large amount of literature is available on the application of interferometry to the study of buoyancy-driven convective fields. Wang and Zhuang [1] used interferometry to evaluate organized structures in high-speed flow past a circular cylinder. Mishra et al [2] experimentally studied the classical Rayleigh-Benard convection problem in an intermediate aspect ratio enclosure with air as the working fluid medium. The optical technique employed was interferometry. Onuma et al [3] carried out a study of crystal growth on microscopic level on Barium Nitrate and K-Alum crystals using schlieren and Mach-Zehnder interferometry. They studied the effect of buoyancy-driven convection and forced flow rate on the microtopography of the crystal growing from its aqueous solution. Tomography for three-dimensional reconstruction of the field of interest can be seen in the work of Faris and Byer [4] for supersonic jets where refraction effects have been accounted for. Synder [5] studied species concentration in a co-flowing jet using tomographic interferometry. Liu et al [6] used speckle photographs to initiate tomographic reconstruction and applied the method to axisymmetric and asymmetric helium jets. Watt and Vest [7] studied structures of turbulent helium jets in air by recording the path integral images based on the refractive index variation using a pulsed phase-shifted interferometer. The advantage of this method is that one can record the phase of the light wave as continuous data, rather than discrete fringes. This greatly improves the spatial resolution of the measurement, being the pixel size rather than the fringe spacing. Subsequently, the authors tomographically reconstructed the helium concentration field. Tomographic measurement techniques and the appropriate reconstruction algorithms suitable for the process industry have been discussed by Mewes et al [8]. Michael and Yang [9] discussed three-dimensional reconstruction of the temperature field using an iterative technique. A Mach-Zehnder interferometer was used by the authors on Rayleigh-Benard convection with water as the working fluid. Mishra et al [10] conducted an interferometric study on Rayleigh-Benard convection using tomography with limited projection data. The working fluid in the experiments was air. Notcovich et al [11] applied interferometric tomography to investigate the three-dimensional temperature field around heavy ice (D_2O) growing from supercooled heavy water. The authors illustrated the use of temperature map in understanding the stability of asymmetrical morphologies which have been observed in ice and other crystals.

The existing literature on the three refractive index-based techniques suggests that only interferometry has been applied extensively for three-dimensional reconstruction of the density fields in the fluid medium. The potential of schlieren technique remains to be explored as a possible tool for quantitative analysis of optically recorded data. The work reported in the present study applies schlieren optical technique for three-dimensional reconstruction of the concentration field around a growing KDP crystal from its aqueous solution.

2. Apparatus and Instrumentation

A schematic diagram of the crystal growth chamber used in the present experiment is shown in Fig. 1. The glass chamber that holds the KDP solution has a diameter of 19 cm and a height of 30 cm. A seed crystal spontaneously crystallized in another vessel is stuck on a thin glass rod, and introduced in the growth cell when the solution temperature is just higher than the saturation value. The diameter of the seed holder is chosen to be as small as possible to ensure minimum disturbance to the convection plumes arising from the crystal surfaces. Circular optical windows (BK-7, 40 mm diameter, 5 mm thickness, $\lambda/4$) are fixed on the glass beaker at opposite ends, in the direction of propagation of the laser beam. Parallelism and straightness of the optical windows are crucial for generating meaningful images, and considerable precautions have been taken in this regard. The Plexiglas tank, $40 \times 40 \times 35$ cm^3 in size, ensures large enough volume for the circulating thermostated water to keep the KDP solution at the necessary temperature as a function of time. Two heating elements placed diametrically opposite in the outer chamber maintain the circulating water, and hence the KDP solution, at the required temperature. Electrical input to the heating element is regulated by a programmable temperature controller (*Eurotherm*). A K-type thermocouple wire fixed to the outer surface of the growth chamber provides the feedback to the controller. Uniformity of temperature within the solution is ascertained by recording temperatures at various locations using 26 gage K-type thermocouples. A linear change in temperature of the solution from 25 to 36°C with time was closely realized in the present set of experiments, over a period of 60 hours.

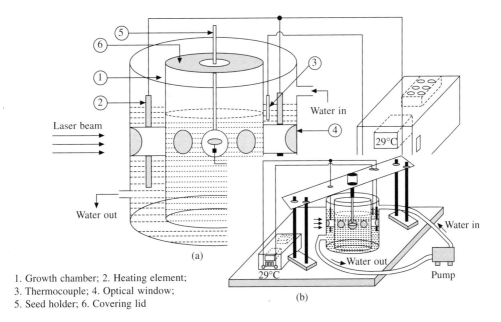

1. Growth chamber; 2. Heating element;
3. Thermocouple; 4. Optical window;
5. Seed holder; 6. Covering lid

Fig. 1 (a) Schematic diagram of the crystal growth chamber and (b) complete assembly.

For optical measurements, a continuous wave helium-neon laser (*Spectra Physics*, 35 mW) has been employed as the coherent light source. A monochrome CCD camera (*Sony*) of spatial resolution of 768 × 574 pixels was used to record the convective field in the form of two-dimensional images. The camera was interfaced with a personal computer (*HCL*, 256 MB RAM, 866 MHz) through a 12-bit A/D card. Light intensity levels were digitized over the range of 0-255. Image acquisition was at video rates of 25 frames/s.

3. Optical Arrangement

The schlieren system used in the present work is of the Z-type as shown in Fig. 2. The optics includes concave mirrors of 1.30 m focal length and 200 mm diameter. Relatively large focal lengths make the schlieren technique sensitive to the concentration gradients [12]. The knife-edge is placed at the focal length of the second concave mirror. It is positioned to cut off a part of the light focused on it, so that in the absence of any optical disturbance, the illumination on the screen is uniformly reduced. The initial intensity values in the experiment were chosen to be less than 20, on a gray scale of 0-255. The knife-edge is set perpendicular to the direction in which the density gradients are to be recorded. In this article, the gradients are expected to be predominantly in the vertical direction parallel to the gravity vector, and the knife-edge has been kept horizontal.

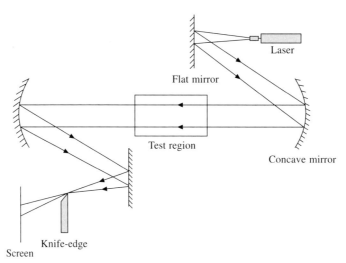

Fig. 2 Schematic diagram of a laser schlieren setup for monitoring convection during crystal growth.

4. Quantitative Analysis

The present section examines the suitability of the schlieren technique for quantitative analysis of the concentration field in the solution during crystal growth. The schlieren arrangement generates projection data, namely information that is integrated in the direction of propagation of the light beam. The result of analysis is thus a concentration distribution that is ray-averaged over the length L (essentially the diameter) of the growth chamber. Extracting the local concentration from the ray-average is possible using principles of tomography [13]. The steps involved in the analysis of the schlieren images are briefly discussed as follows.

Refractive index techniques depend on the fact that for a transparent material, refractive index and density have a unique relationship. In addition, density is proportional to solute concentration and temperature. In the present analysis, it is assumed that the aqueous solution in the crystal growth chamber has negligible temperature gradients. The assumption is reasonable because the solution is cooled very slowly and is close to thermal equilibrium throughout the experiment. Hence, refractive index becomes a measure of concentration itself. The material property that determines the sensitivity of the optical measurement is dn/dC, where n is the refractive index and C the solute concentration.

Image formation in a schlieren system is due to the deflection of light beam in a variable n-field towards regions of higher refractive indices. In order to recover quantitative information from a schlieren image, one has to determine the deflection angle as a function of position in the x-y plane that is normal to the light beam. Using principles of ray-optics, the angular deflection α can be expressed as

$$\alpha = \frac{1}{n_0} \int_0^L \frac{\partial (\ln n) n}{\partial y} dz$$

The intensity field can now be related to the refractive index field directly as

$$\frac{\Delta I}{I_k} = \frac{f}{a_k n_a} \int_0^L \frac{\partial n}{\partial y} dz$$

where n_a (~1) is the refractive index of the ambient. In terms of the ray-averaged refractive index, the governing equation for the schlieren process is derived as

$$\frac{\Delta I}{I_k} = \frac{f}{a_k} \frac{\partial n}{\partial y} L \qquad (1)$$

Eq. (1) requires the approximation that changes in the light intensity that occur due to beam deflection, rather than its physical displacement.

4.1 Computerized Tomography

The three-dimensional concentration field around the growing crystal can be reconstructed from its projection data using the principles of tomography. Tomography is the process of recovery of a function from a set of its line integrals evaluated along some well-defined directions. The projection data is recorded using the parallel beam geometry in the present work. Tomographic algorithms used in the present work reconstruct two-dimensional concentration fields from their one-dimensional projections. Reconstruction is then applied sequentially from one plane to the next until the third dimension is filled.

Tomography can be classified into three major categories: (a) transform, (b) series expansion, and (c) optimization methods. *Transform* methods generally require a large number of projection data for a meaningful answer. The convolution back projection (CBP) algorithm for three-dimensional reconstruction can be classified into this category. For the past several decades, CBP has been used for medical imaging of the human brain. Significant advantages of this method include its non-iterative character, availability of analytical results on convergence of the solution with respect to the projection data, and established error estimates. The use of CBP continues to be seen in steady flow experiments, particularly when the region is physically small. *Series expansion* methods are iterative in nature and are the most appropriate for applications where limited projection data is available from experiments. Algebraic reconstruction techniques

(ART) can be classified under the series expansion methods category. *Optimization-based* algorithms are known to be independent of initial guess, but the choice of the optimization functional plays an important role in the result obtained.

In practice, projection data can be recorded either by turning the experimental setup or the source-detector combination. In the present study, the latter is particularly difficult due to stringent requirement of alignment. With the first option, it is not possible to record a large number of projections, owing to the inconvenience of installing plane optical windows in a circular beaker. Hence, with the limited available data, the ART family of reconstruction algorithms has been employed to reconstruct the concentration field in the vicinity of the growing crystal. For a cylindrical crystal growth chamber, the entire field of interest cannot be imaged due to the curvature of the test cell. Instead, the central core region (40 mm diameter), that includes the growing crystal, has been recorded. Clearly, the projection data set is incomplete. In order to generate a complete projection data set for each view angle, the following approach has been adopted:

1. For a given view angle, record the partial projection data.
2. Determine the ray-averaged refractive index field on selected planes by integrating Eq. (1).
3. Extrapolate the data suitably to fill the diameter of the growth chamber. At this stage, the continuity requirements of the refractive index field should be ensured.
4. Apply steps 1-4 for all the planes.
5. Apply 1-5 for all the view angles.

Projection data has been recorded using four view angles (0°, 45°, 90° and 135°) in the present experiments. Information about the intermediate view angles has been generated by employing linear interpolation on the experimentally recorded projection data. The algorithms used in the present work have been adapted from [13]. The present treatment of reconstruction with partial data is similar to that of [14].

5. Results and Discussion

The time sequence of the schlieren images, and images recorded at various view angles are reported in the present section. The ray-averaged refractive index field and the reconstructed field on selected planes below the crystal are also presented.

5.1 Convection Around a Growing KDP Crystal

When a KDP seed is inserted in the aqueous solution, the instantaneous temperature difference between the two leads to the dissolution of the seed. The local density of the solution increases, and consequently the solution descends vertically from the crystal. With the passage of time, thermal equilibrium is established, and density differences within the solution are solely due to concentration differences. Adjacent to the crystal, the solute deposits on the crystal faces, and the solution goes from the supersaturated to the saturated state. Thus, the solution near the crystal is lighter than the solution away from it. The denser solution displaces the lighter fluid, and a circulation pattern is set up around the crystal. The structured movement of the fluid, called a buoyant plume is essential for transporting the solute from the bulk of the solution to the crystal and determines the crystal growth rate. The plume is visible as the spread of light intensity in the schlieren images.

Fig. 3 shows the growth sequence as recorded using the schlieren technique. The first image

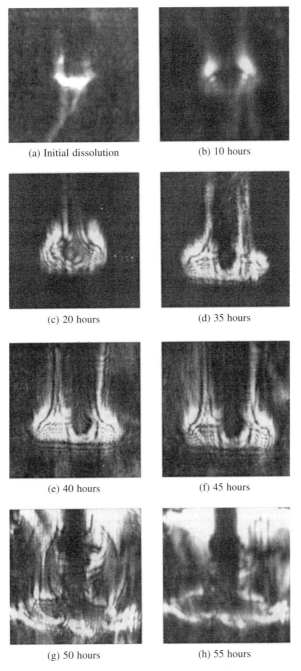

(a) Initial dissolution (b) 10 hours
(c) 20 hours (d) 35 hours
(e) 40 hours (f) 45 hours
(g) 50 hours (h) 55 hours

Fig. 3 Schlieren images of the transient evolution of the convective field around a KDP crystal growing from its aqueous solution.

shows initial dissolution of the seed just after its insertion into the solution. A sharp descending plume originating from the seed can be seen. The intensity contrast is related to an abrupt change in the solute concentration around the seed crystal, which creates a jump in the refractive index, and deflects the light beam into the region of relatively large concentration gradients. After the initial dissolution, the growth process of the crystal is initiated. The associated convection currents are larger, and a significant increase in the size of the bright region is seen. With the progress of the growth process, the schlieren images show the upward movement of the buoyant plumes around the crystal. These plumes end in the bulk of the solution in the beaker, and descend in such a way as to form a closed loop. These are responsible for the deposition of the solute on the crystal surface. The plumes are almost steady and uniform in nature resulting in uniform and symmetric growth of the crystal. This phase of the growth process can be identified as the stable growth regime in which the associated flow field is highly symmetric and delivers high crystal quality. The gradient of concentration is larger near the crystal surface when compared to the bulk of the solution. This distribution can be visualized as a diffusion boundary layer around the growing faces of the crystal. The thickness of the boundary-layer can be identified as the region over which the light intensity is high. The vertically upward movement of the convective plumes results in the variation of the thickness of concentration boundary layer along the crystal faces. During the growth phase of the crystal, the boundary-layer is the thinnest on the lower side of the crystal, and the highest on the upper side. As the crystal size exceeds a critical value, the flow field approaches a stagnant condition when the solution becomes stratified. The layering and stable stratification of the solution are seen for times greater than 50 hours (Fig. 3 (g, h)) and the corresponding schlieren images show the shift of brightness in the regions away from the growing crystal.

5.2 Quantitative Analysis

The present section discusses the suitability of the laser schlieren technique as a tool for quantitative analysis of the concentration field around the growing crystal and its three-dimensional reconstruction using tomographic algorithms. Results have been discussed in the form of two-dimensional concentration contours in the vicinity of the growing crystal, concentration profiles and concentration field at various horizontal planes in the growth chamber.

5.2.1 Concentration Contours

Figure 4 shows concentration contours extracted from the schlieren images. The central vertical line represents the position of the seed holder on which the growing crystal is fixed and is the origin of the x-coordinate. The horizontal mid-plane of the crystal is at a location $y = 0$. The x and y coordinates are non-dimensionalized with respect to the maximum crystal size grown in the experiment. The concentration values have been normalized to the range $0 - 1$, $C = 0$ representing the saturated state, and $C = 1$ being the supersaturated condition at the temperature of the growth chamber. The local variation of concentration with respect to the vertical coordinate is presented in Fig. 5. Four columns placed symmetrically on either side of the seed holder have been chosen in this regard. The contours in the figure indicate an almost uniform distribution of solute in the growth chamber, both near the crystal and in the bulk of the solution during the early stages of the experiments ($t < 10$ hrs). Nearly overlapping concentration profiles in Fig. 5 during this period of growth also reveal uniform concentration distribution. Uniform incorporation of solute on the growing surfaces is seen, resulting in symmetric concentration profiles on either side of the crystal. With the passage of time, the crystal size increases and the gradients increase

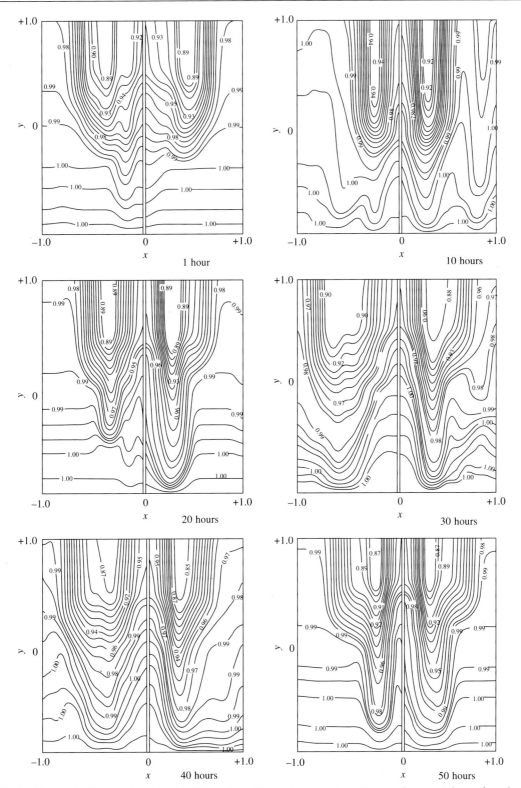

Fig. 4 Concentration contours derived from the schlieren images above the growing crystal, as a function of time. A concentration of unity shows that the solution is supersaturated.

in strength. While the intensity fields in the schlieren images of Fig. 3 look broadly symmetric, an asymmetry in the concentration distribution is brought out by the detailed quantitative analysis. This is evident from the differences in the local concentration in the bulk of the solution at $t = 30$ hours (Fig. 5). A higher density of concentration contours in Fig. 4 near the crystal-solution interface indicates the presence of high gradients because of the transport of solute from the solution to the growing surfaces.

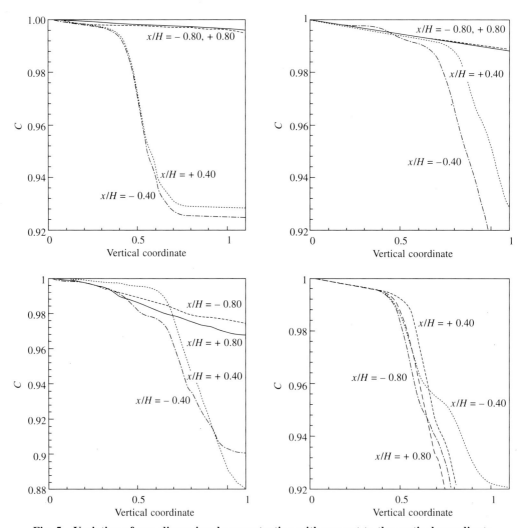

Fig. 5 Variation of non-dimensional concentration with respect to the vertical coordinate.

5.2.2 Three-dimensional Reconstruction

The concentration data available on a Cartesian grid for each view angle, derived from the schlieren images, represents the line integral of the concentration field in the fluid layer.

Concentration fields on selected planes below the crystal have been reconstructed using tomographic algorithms, namely CBP and combined ART-CBP. In the experiments, schlieren images have been recorded using 4 view angles (0°, 45°, 90° and 135°). Fig. 6 shows the schlieren images recorded at a given time instant from 4 view angles. They have been utilized to reconstruct the concentration field on various planes of the aqueous solution. The region below the growing crystal has been chosen for analysis. Concentration fields have been reconstructed on four horizontal planes (y/H = 0.10, 0.20, 0.30 and 0.40, H being the height of the growth chamber). Ray-averaged concentration contours corresponding to the images of Fig. 6 are shown in Fig. 7(a) for all the four view angles. Broadly, the contours predict a symmetric concentration field in the vicinity of the growing crystal around the beaker axis with a slight disturbance in the bulk of the solution, in particular near the boundaries. It is to be mentioned here that these concentration contours correspond to the stable regime of the growth process. Ray-averaged

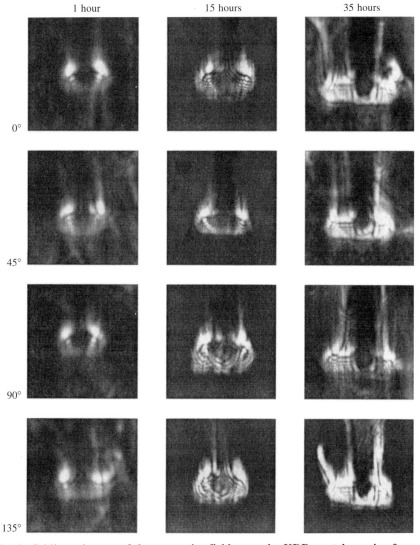

Fig. 6 Schlieren images of the convective field around a KDP crystal growing from its aqueous solution as recorded from the four view angles (0°, 45°, 90° and 135°).

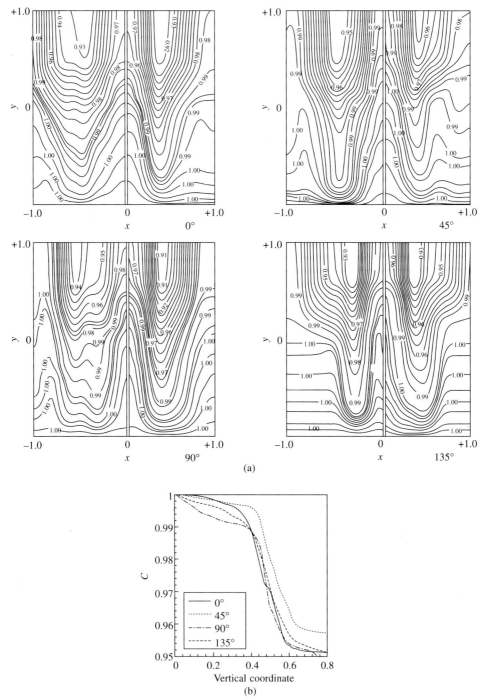

Fig. 7 (a) Concentration contours derived from the schlieren images around the growing crystal as a function of projection angle and (b) Width-averaged concentration profiles with respect to the vertical coordinate for the four view angles.

concentration profiles of Fig. 7(a) show that the concentration field as viewed from the four view angles are quite similar. The curves almost overlap in the region below the growing crystal ($y/H < 0.40$), while differences are noticeable in the upper half of the solution. The disturbance in the upper half can be partly attributed to the presence of the seed holder (with its axis

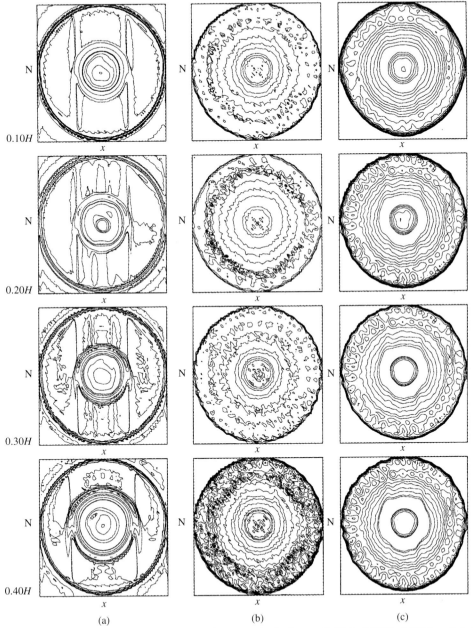

Fig. 8 Reconstructed concentration field on four vertical planes (0.10H, 0.20H, 0.30H and 0.40H, H being the height of the growth chamber). Tomographic algorithm employed; (a) CBP, (b) ART and (c) combined ART-CBP.

coinciding with the axis of the beaker), that affects the movement of the buoyant plumes rising above the growing crystal. Similar conclusions can be drawn from the average concentration profile plotted with respect to the vertical coordinate in Fig. 7(b). The closeness of the profiles for the individual view angles is, in fact, a mass balance check on the experimental data.

Fig. 8 shows reconstructed concentration contours over four horizontal planes of the fluid layer. Fig. 8(a) shows the reconstruction with CBP as the tomography algorithm. The reconstruction predicts almost symmetric field (in the form of concentric circles) about the beaker axis in the central region. Unphysical artifacts are, however, seen in the periphery, indicating that CBP is not an appropriate algorithm of choice for partial projection data. Fig. 8 (b, c) shows the reconstructed field when the ART algorithm is used, with the CBP reconstruction as the initial guess. Fig. 8(b) has been generated with the output of the ART algorithm, while Fig. 8(c) is the output of the CBP algorithm when the projection data is based on Fig. 8(b). In comparison to the reconstruction obtained with CBP alone, the combined approach reduces the spurious artifacts, and predicts an almost symmetric concentration field in the entire growth chamber.

A closer analysis of the reconstruction techniques and errors for a wider range of experiments will form the topic of future research.

6. Conclusions

An investigation of the buoyancy-driven convective field in the vicinity of a growing KDP crystal has been conducted using a monochrome schlieren technique. The three-dimensional concentration field around the crystal is determined from the schlieren images using principles of tomography. Tomography algorithms employed are CBP and combined ART-CBP. Qualitatively, the growth process shows quite clearly the spatial distribution of concentration differences in the vicinity of the growing crystal. The initial dissolution and unsteadiness, a stable and symmetric growth regime and the final stratification of the solution are brought out. Quantitative analysis of the schlieren images reveals the finer aspects of the flow field in terms of unsteadiness and asymmetry in the growth chamber in the initial stages of the growth process.

The concentration field reconstructed on four different horizontal planes of the flow field reveals the presence of an overall symmetry around the beaker axis, with the concentration contours forming concentric rings. The data reconstructed with CBP shows unphysical artifacts, indicating that the algorithm is unsuitable for partial and limited projection data. When used as an initial guess to the ART algorithm, the results show an improvement and the symmetry of the projection data is seen in the local concentration contours on selected planes. The reconstruction is, however, noisy, and points towards the need for a more systematic research.

References

1. Wang D. and Zhuang T., The measurement of 3-D asymmetric temperature field by using real time laser interferometric tomography, *Optics and Lasers in Engineering* 2001; **36**: 289–297.
2. Mishra D., Muralidhar K. and Munshi P., Interferometric study of Rayleigh-Benard convection at intermediate Rayleigh numbers, *Fluid Dynamics Research* 1999; **25(5)**: 231–255.
3. Onuma K., Tsukamoto T. and Nakadate S., Application of real time phase shift interferometer to the measurement of concentration field, *Journal of Crystal Growth* 1993; **129**: 706–718.
4. Faris G.W. and Byer R.L., Three-dimensional beam deflection optical tomography of a supersonic jet, *Applied Optics* 1988; **27(24)**: 5202–5212.

5. Synder R., Instantaneous three-dimensional optical tomographic measurements of species concentration in a co-flowing jet, Report No. SUDAAR 567, Stanford University, USA 1988.
6. Liu T.C., Merzkirch W. and Oberstr-Lehn K., Optical tomography applied to speckle photographic measurement of asymmetric flows with variable density, *Experiments in Fluids* 1989; **7**: 157–163.
7. Watt D.W. and Vest C.M., Turbulent flow visualization by interferometric integral imaging and computed tomography, *Experiments in Fluids* 1990; **8**: 301–311.
8. Mewes D., Friedrich M., Haarde W. and Ostenford W., Tomographic measurement techniques for process engineering studies, *Handbook of Heat and Mass Transfer* 1990; 3 (Chapter 24).
9. Michael Y.C. and Yang K.T., Three-dimensional Mach-Zehnder interferometric tomography of the Rayleigh-Benard problem, *ASME J. Heat Transfer* 1992; **114**: 622–629.
10. Mishra D., Muralidhar K. and Munshi P., Interferometric study of Rayleigh-Benard convection using tomography with limited projection data, *Experimental Heat Transfer* 1999; **12(2)**: 117–136.
11. Notcovich A.G., Braslavsky I. and Lipson S.G., Imaging fields around growing crystals, *Journal of Crystal Growth* 1999; **198/199**: 10–16.
12. G.S. Settles, Schlieren and shadowgraph techniques: Visualizing phenomena in transport media, Springer-Verlag, New York, 2001.
13. Mishra D., Muralidhar K. and Munshi P., A robust MART algorithm for tomographic applications, *Numerical Heat Transfer B (Fundamentals)* 1999; **35(4)**: 485–506.
14. Mishra D., Longtin J.P., Singh R.P. and Prasad V., Performance evaluation of iterative tomography algorithms for incomplete projection data, *Applied Optics* 2004; **43(7)**: 1–11.

Computerized Tomography for Scientists and Engineers
Edited by P. Munshi
Anamaya Publishers, New Delhi, India

13. Development of Computer Aided Tomography Systems in DRDL

S. Vathsal, C. Muralidhar, G.V. Siva Rao, K. Kumaran, M.P. Subramanian, M.R. Vijaya Lakshmi, Sijo N. Lukose and M. Venkata Reddy

Non-Destructive Evaluation Division, Defence Research & Development Laboratory (DRDL), Kanchanbagh, Hyderabad-500 058, India

Abstract

Computer aided tomography (CAT) system has been developed with 450 kV X-ray source, 256-channel detector array with four-axes mechanical manipulator for scanning objects of 300 mm dia and 200 kg. Another CAT system, which is under integration uses a six-axes mechanical manipulator for 1000 mm diameter and 2000 kg objects. Major aspects of four-axes system development are highly precise mechanical manipulator, alignment of source and detector, electrical integration, electromagnetic compatibility and synchronized data acquisition system. Scanning modes of rotate only, translate-rotate were implemented to accommodate different sizes of objects and to achieve desired resolution. A new method of scanning the object by sensing the values was adopted for translate-rotate scan. Filtered back projection (FBP) has been used as reconstruction algorithm and the spatial resolution achieved is 1-line pair/mm. Corrections for photon statistics, center of rotation error, ring and streak artifacts were implemented. System characterization studies were carried out by plotting contrast against defect size. Pre-hardening method was used for correcting beam hardening effect. Different objects with varying densities were scanned and analyzed. Region of interest (ROI) reconstruction and 3D rendering were implemented for critical defects analysis. Various image-processing techniques were employed for better visualization and interpretation.

1. Introduction

Computed tomography (CT) generates 2D cross-sectional images of the object from 1D measurement data and the CT image represents point-by-point linear attenuation coefficients in the slice [1, 2]. CT images are free from underlying and overlying areas of the object. Tomography system basically consists of radiation source, detector, mechanical manipulator and reconstruction software.

Development of CAT system is multidisciplinary in nature as it involves expertise from nondestructive testing, mechanical, electrical, electronics, computer science and material science backgrounds. The objects that are scanned using CAT system vary widely in their shape, size, complexity and density. Though medical CT can be used for scanning certain objects, it has limitations for energy selection, gantry size and time of scanning to accommodate different industrial objects. Similarly, gamma-ray CT system has limitation for energy selection due to discrete energy levels available, lower intensities, bigger focal spot, and longer scanning times as compared to X-ray CT. Due to the above reasons, X-ray based CAT systems have been developed to meet our nondestructive evaluation (NDE) requirements.

This article discusses various aspects evolved during the development of CAT:

(i) Design and development of mechanical manipulators for handling large industrial objects.
(ii) Alignment and integration problems.
(iii) Scanning modes for accommodation of different objects (size and shape).
(iv) Various artifacts and their correction.
(v) Study of contrast vs defect size.
(vi) Development of methods for beam hardening correction.
(vii) Post-processing techniques for extracting details from the image and better visualization using 3D rendering.
(viii) Integration of sub-systems with software.

2. Developmental Aspects

2.1 Design and Development of Mechanical Manipulators

The 450 kV X-ray source with 1.8 mm and 3.6 mm focal spot sizes, 256-channel detector array of 18 bit dynamic range having 0.8 mm aperture and 1.3 mm pitch were used with four and six-axes mechanical manipulators.

Four-axes: This is a prototype manipulator developed for handling objects upto 300 mm diameter and 200 kg weight with X-, Y-, Z-, C-axes to accomplish rotate-only and translate-rotate configurations with an accuracy of 25 microns. Source and detector are stationary and the object height is manipulated for selecting the slice. The angular and linear resolution of the scanner is 0.001° and 0.001 mm. Fig. 1 shows the CT system with four-axes mechanical manipulator along with X-ray source and detector array.

Six-axes: It is developed for handling objects upto 1000 mm diameter and 2 tonnes weight with X-, Y_1-, Y_2-, Z_1-, Z_2-, C-axes. Source and detector are integral part of the system and can be moved synchronously for selecting the slices. They are also provided with six degrees of freedom for precise alignment. The angular and linear resolution of the scanner is 0.001° and 0.001 mm. Rotate-only and translate-rotate configurations were used to accommodate different types of objects and desired resolutions. Fig. 2 shows six-axes manipulator based CT system.

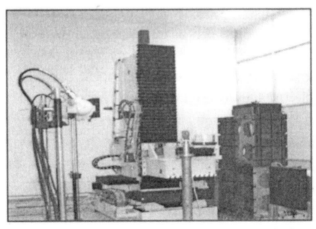

Fig. 1 The CT system with four-axes mechanical manipulator along with X-ray source and detector array.

Fig. 2   The six-axes manipulator based CT system.

2.2   Alignment and Integration Problems

Precise mechanical alignment is the primary requirement of a CT system. Coplanarity and collinearity were established between the source, detector and rotary table. The overall alignment was within an accuracy of 50 microns. Severe electromagnetic interference was affecting the detector data. Conductive interference was corrected by putting a matching line filter with the manipulator. Radiative interference was also corrected by using twisted pair shielded cables.

2.3   Data Acquisition Systems

Triggering of the data acquisition system is possible by sensing the manipulator positions. This feature has been used to collect data samples during object motion in a synchronous manner in rotate-only and translate-rotate configurations.

2.4   Scanning Geometries

The scanning configurations (rotate-only and translate-rotate) were selected to accommodate different sizes of objects and desired resolutions [3]. In rotate-only geometry objects only upto 220 mm diameter can be scanned due to the limitation of X-ray fan angle of 10°. Direct fan beam reconstruction is implemented using modified filtered back projection (FBP) algorithm. A resolution of 0.6 lp/mm was achieved in rotate-only configuration.

In conventional industrial translate-rotate computed tomography systems, the X-ray source and detector array are stationary and the object is linearly traversed across the beam to make a series of X-ray transmission measurements from multiple detectors. While scanning a circular object as the object moves in the linear axis, the time for which the object remains within a detector (dwell time) varies for the extreme rays in the fan beam compared to those at center due to different angle of intersection of the rays with the object. The variation of ray spacing from central ray to extreme ray is shown in Fig. 3. Due to this, one side of the sinogram is affected and manifests as artifact in one-half of the tomogram as shown in Fig. 4. To correct for this artifact, the object is traversed in an arc path in such a way to move the object in a path, whose slope is perpendicular to all the rays in the beam so that the resulting parallel projections will all

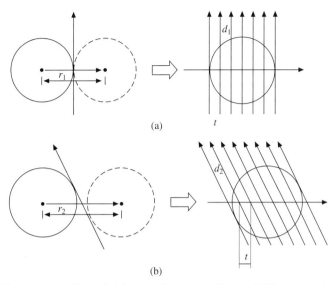

Fig. 3 Schematic representation of: (a) central ray sampling and (b) non-central ray sampling.

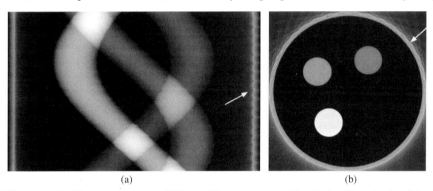

Fig. 4 (a) Sinogram for linear traverse and (b) resulting tomogram (the chain of arcs in the right side of the sinogram and the resulting artifacts in the upper portion of the tomogram are indicated by the arrows).

have the same ray spacing. Hence, after each translation the object is rotated by fan angle. The above procedure is repeated to obtain data over 180°. In this way the data collected at different view angles will have equal ray spacing and the artifact manifested due to unequal ray spacing is corrected as shown in Fig. 5. Parallel projection data for full 180° is sorted out from fan beam data obtained during the translation motions [4]. A resolution of 1 lp/mm was achieved in translate-rotate scanning configuration.

Figure 6 shows the resolution phantom revealing 0.6 lp/mm resolution in rotate-only configuration, whereas 1 lp/mm in translate-rotate configuration due to finer radial sampling.

3. Results and Discussion

3.1 Various Artifacts and Their Correction

Corrections for photon statistics, center of rotation error, ring and streak artifacts were incorporated before reconstruction. Image processing operations such as contrast enhancement, edge detection,

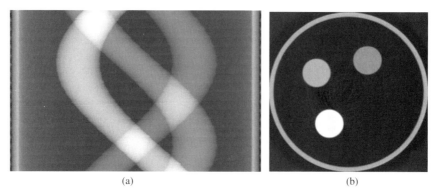

Fig. 5 (a) Sinogram for arc traverse and (b) resulting tomogram.

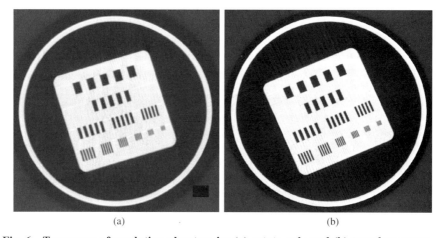

Fig. 6 Tomogram of resolution phantom in: (a) rotate-only and (b) translate-rotate.

pseudo coloring, dimensioning and filtering were carried out as post-processing operations on tomograms for defect analysis. Pre-hardening method (placing filters at X-ray source) was used for correcting the beam hardening effect, 3D visualization and region of interest (ROI) reconstruction were incorporated for defect characterization.

3.2 Beam Hardening Effect and Its Correction Using Filters

X-ray source is poly-energetic (photons of different energies) in nature and the attenuation at a point is generally greater for photons of lower energy, the energy distribution (spectrum) of X-ray changes (hardens) as it passes through the object. This is known as beam hardening effect. The correction for beam hardening was done using filters of various materials (copper, steel etc.) of varying thickness placed at the X-ray source.

Figure 7 shows the tomograms of perspex cylinders of 50 mm diameter. It is clear from the density profiles that the cupping artifact is not observed and the beam hardening effect is not significant.

Figure 8 shows the tomogram of aluminum cylindrical rod of 50 mm diameter. The cupping artifact is clearly seen and corrected by 2.1 mm of copper filter. Equivalent thickness of steel

Development of Computer Aided Tomography Systems in DRDL 153

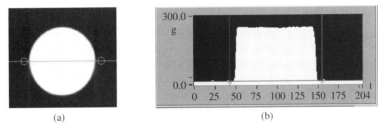

Fig. 7 Tomogram of 50 mm diameter perspex cylinder: (a) image and (b) density profile.

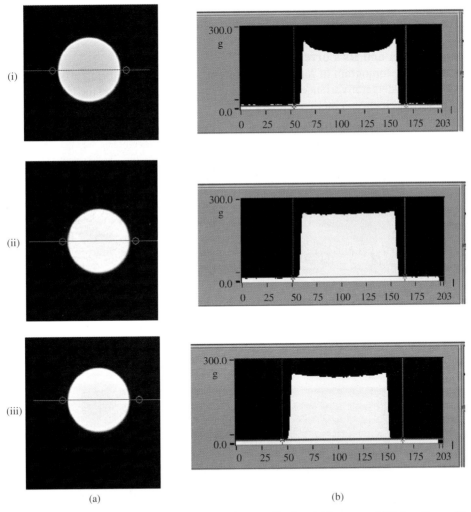

Fig. 8 Tomogram of 50 mm diameter aluminum cylinder: (a) image and (b) density profile. (i) Without beam hardening correction; (ii) corrected with 2.1 mm copper filter and (iii) corrected with 2.4 mm steel filter.

(2.4 mm) in terms of copper (2.1 mm) is also used as a filter and found that the extent of beam hardening correction done in both cases is not the same.

The correction for beam hardening effect using pre-hardening method is effective as is evident from the results discussed above. However, the exact filter thickness required may slightly vary with the type of energy spectrum of the source used. The beam hardening correction with filters at the source often provides some degree of compensation for scatter also by cutting off the lower energy photons, which contribute for scattering.

It is observed from our studies that the beam hardening effect is not significant in low dense materials [5] (perspex) even if the diameter is large enough (215 mm), whereas the on-set starts with Teflon (high dense material) and is predominant in highly dense materials (aluminum) of smaller diameter (50 mm) and is corrected using 2.1 mm copper filter. It is evident from our studies that the beam hardening correction is required for solid objects of densities greater than 2 g/cc.

3.3 Ring Artifact and Its Correction

Figure 9 shows the tomogram of resolution phantom with and without ring artifacts. These ring artifacts are due to detector variance, which were corrected by increasing the transmittance of X-rays and ensuring electromagnetic compatibility.

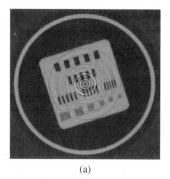 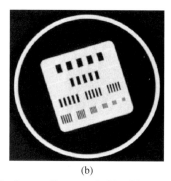

(a)   (b)

Fig. 9 Tomogram of resolution phantom: (a) with ring artifacts and (b) without ring artifacts.

The tomograms of contrast phantom with and without (corrected) center of rotation error are shown in Fig. 10. Any deviation of center of rotary table from the reconstruction center results in artifacts in the image, which is corrected by estimating the center of rotation from the sinogram.

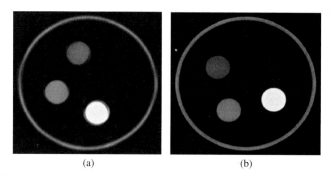

Fig. 10 Tomogram of contrast phantom for center of rotation error correction: (a) without correction and (b) with correction.

3.4 Contrast vs Defect Size

Contrast in computed tomography (CT) can be expressed as percentage difference of a feature from a background material [6]. Fig.11 shows the contrast vs defect size for the uniform aluminum cylinders having different defect sizes (0.4-3.6 mm) before and after beam hardening correction. It is clear from the figure that contrast increases with the defect size and tends to saturate thereafter. Features (defect) of size less than minimum beam width cannot be resolved with reasonable contrast. A defect of 0.4 mm size was resolved with minimal contrast and the contrast is improved after beam hardening correction.

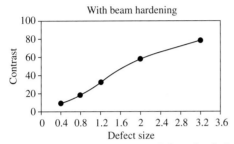
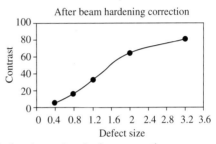

Fig. 11   Contrast vs defect size before and after beam hardening correction.

X-ray beam profile varies between the source and detector and affects the contrast and resolution of the CT image. A cylinder having defect size of 2 mm was selected for studying the variation of contrast by placing the object at different positions between source and detector. The contrast vs position of object is shown in Fig. 12. The results revealed that the contrast is poor near the source, improves gradually and reaches maximum at the minimum beam width position [7, 8]. It decreases slightly thereafter and remains almost constant towards the detector because the rate of divergence of the radiation beam is less towards detector when compared towards source.

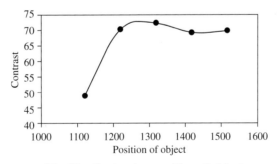

Fig. 12   Contrast vs position of object.

Fig. 13 shows the tomograms of FRP casing, composite casing and ceramic radome revealing various types of defects such as delaminations, debonds and cracks.

The 3D visualization of an object from the 2D slices and its cut-out view is shown in Fig. 14.

4. Conclusion

X-ray CT systems have been developed with four-axes and six-axes mechanical manipulators to cater for our requirements. Rotate-only and translate-rotate scanning geometries were selected to accommodate different sizes of objects. Corrections for photon statistics, center of rotation

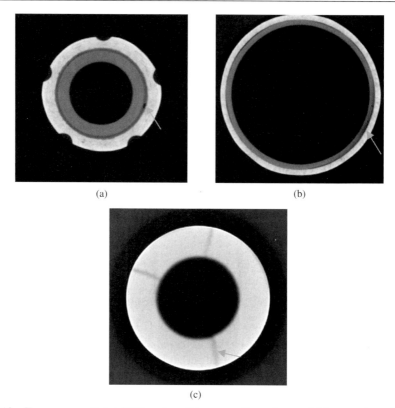

Fig. 13 Tomograms of: (a) FRP casing, (b) composite casing and (c) ceramic radome.

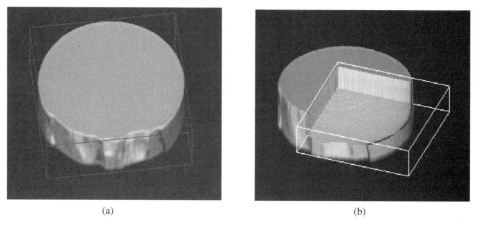

Fig. 14 (a) The 3D visualization of an object from its 2D slice images and (b) cut-out view of the same.

error, ring and streak artifacts were implemented. Pre-hardening method was used for correcting the beam hardening effect. Noise of system (standard deviation) was experimentally found to be within 1%. Different types of objects with varying sizes, densities have been scanned and analyzed. The 3D rendering has been implemented for the critical NDE analysis.

Acknowledgements

The authors wish to express sincere thanks to Shri Prahlada, Director, DRDL for support and constant encouragement. They express thanks to Shri I.K. Kaul (Scientist G), Director, DOE, Shri P.K. Biswas (Scientist G) (Rtd), Shri S.B. Kumar (Scientist F) (Rtd), Shri M. Damodar (Scientist G), Shri P. Rajeswara Rao (Scientist G), Prof. K. Rajagopal, IISc and Prof. P. Munshi, IIT Kanpur, for their support during development phase.

References

1. Avinash C. Kak, Malcolm Slaney, "Principles of Computerized Tomographic Imaging", IEEE PRESS, 1988.
2. Gabor T. Herman, "Image Reconstruction from Projections: the Fundamentals of Computerized Tomography", Academic Press Inc. 1980.
3. Michael J. Dennis, "Industrial Computed Tomography", ASM Handbook, Non-Destructive Evaluation and Quality Control, Volume 17, 358–385, May 1994.
4. S. Chakraborty, Sijo N. Lukose and M.P. Subramanian, "Implementation Problems and Solutions in Multi-detector Translate-Rotate Tomography system", Proceedings of NDE 2003.
5. C. Muralidhar, K. Kumaran, M.R. Vijaya Lakshmi, "Beam Hardening Effect and Its Correction Using Filters in Computed Tomography", Proceedings of NDE 2003.
6. Paul Burstein, "Performance Issues in Computed Tomography Specifications", Materials Evaluation, 579–593, May 1990.
7. Simon H.C. Hughes and Lowell D. Harris, "Utilizing the non-uniform spatial resolution of a CT image by correct positioning of the half power cross-over point", ASNT Topical Proceedings, Seattle, W.A, July 25–27, 1989.
8. G.V. Siva Rao, M.R. Vijaya Lakshmi and Sijo N. Lukose "A Study of Contrast vs Defect Size in Computed Tomographic Imaging", Proceedings of NDE 2003.

Computerized Tomography for Scientists and Engineers
Edited by P. Munshi
Anamaya Publishers, New Delhi, India

14. Determination of the Concentration Field Around a Growing Crystal Using Laser Shadowgraphic Tomography

Sunil Verma, K. Muralidhar* and V.K. Wadhawan

Laser Materials Division, Centre for Advanced Technology, Indore-452 013, India
*Department of Mechanical Engineering, Indian Institute of Technology, Kanpur-208 016, India

Abstract

Growth rate and quality of a KDP crystal grown from its aqueous solution are strongly influenced by the nature of the convection pattern in the liquid. We have visualized buoyancy driven convection in the aqueous solution using the laser shadowgraph technique. The steps involved in obtaining the three-dimensional concentration field using principles of computerized tomography are discussed. Shadowgraph images show that the strength of the convection currents gradually increases with the crystal size. The plume behaviour is initially laminar, becoming unsteady subsequently, and finally becoming chaotic after 190 h of growth. Shadowgraph images along 19 angular directions have been captured. They constitute the projection data for the 3D tomographic reconstruction of the concentration field above the crystal surface. The experimental projection data is converted to equivalent refractive index maps by numerically solving the governing equation of the shadowgraph process. Both, the *convolution back projection* as well as the *iterative algorithm* for tomographic reconstruction are discussed and implemented for 3D reconstruction of the concentration field from 2D projection data. The experimental results reported in the present work show that shadowgraphic tomography is a useful technique for monitoring crystal growth, and for revealing the three-dimensional nature of the solutal concentration field.

1. Introduction

Crystals are needed for various advanced technological applications, such as laser host application, frequency conversion, electro-optic switching, photorefractive application, radiation detection, holographic data storage and semiconductor application, to name a few. The requirement on the quality of crystals for these applications is very high, and is constantly a moving target for crystal growers. The aim of a crystal grower is to grow as big a crystal as possible, while maintaining the crystal quality desired for the end application. This requires a careful optimisation of several growth influencing parameters operating simultaneously. Transport phenomenon, which includes heat and mass transfer, and fluid flow, has been found to strongly influence the growth process. In particular, crystal growth from solution, which involves double diffusive convection, is intricately affected by the transport conditions. Depending upon the mode of convection adopted, the crystallization process operates either in the *diffusion regime* or in the *kinetic regime*. During free convection conditions the process operates under the former regime, and it shifts to the latter when forced flow is adopted. The quality of crystal grown under diffusion regime is generally good, but the growth rate is small so as to make the use of diffusion regime conditions un-exploitable for commercial reasons. In particular, to grow those crystals that are needed in large dimensions, such as the size of KDP plates needed for inertial confinement fusion experiments, it becomes a necessity to adopt forced flow conditions. Growth under forced flow is a tough optimisation problem, because the associated deleterious effects such as morphological instability, spurious nucleation, and inclusion incorporation, have increased probability of occurrence under unsteady or turbulent flow dynamics. The role of transport phenomenon in crystal growth from solution has been discussed in a few reports [1-3]. It affects the compositional homogeneity, surface microstructure and the growth rate of the crystal. The importance of convection has been examined by three-dimensional time-dependent simulations of fluid flow and mass transfer around KDP and KTP crystals [4-8], and by the work of Zaitseva and co-workers at LLNL, USA in the context of rapid growth of KDP crystals [9].

Optical techniques have seen resurgence in the engineering research, particularly in the field of experimental fluid mechanics because of the availability of lasers as sources and computers for data acquisition and processing. They have the specific advantage of being non-intrusive in nature, and provide inertia-free and whole-field measurement of the process under study. The optical techniques of interferometry, schlieren, shadowgraphy, holography, laser Doppler velocimetry, particle image velocimetry, speckle and Moiré interferometry and scattering based techniques [10-13] provide a wide canvas for studying a variety of engineering processes, on the small as well as the large scales. The only limitation of the optical techniques is that information of the three-dimensional processes is available as two-dimensional images. However, this limitation can be overcome by using the principles of computerized tomography [14]. Here several two-dimensional images of the process are taken at different view angles, followed by the application of an appropriate reconstruction algorithm to obtain the three-dimensional structure of the field parameter, such as solute concentration.

In order to study the relationship between convection and the growing crystal, we have used the laser shadowgraph technique to visualize the convective plumes rising above the growing crystal. The growth chamber is maintained under practically isothermal condition. Convection arises from the difference in concentration of KDP in water (and hence the solution density) between the region adjacent to the crystal and the bulk of the solution. Since thermal gradients are much weaker when compared to the concentration gradients driving the convective process, we have determined the concentration maps above the surface of the growing crystal at varying heights. Nineteen shadowgraph images have been recorded at an angular interval of $10°$. These images constitute the experimental projection data. The images are converted to refractive index field (and hence the concentration field) by solving the governing equation of the shadowgraph process [15]. The resulting two-dimensional refractive index field constitutes the numerical projection data required for tomographic reconstruction [16-19].

Section 2 describes the principle of the shadowgraph optical technique and the mathematical formulation that leads to the governing equation of the shadowgraph process. Section 3 presents the details of the numerical procedure and the computer codes developed to solve the governing equation. Section 4 is a summary of principles of tomography and the conditions under which tomographic reconstruction is possible in the context of crystal growth. Section 5 is a description of the crystal growth apparatus, the temperature controlling circuitry, and the image acquisition and data processing hardware. Section 6 discusses the crystal growth procedure adopted in the present work, while Section 7 describes the shadowgraph experiments performed. Section 8 presents the results of shadowgraph imaging and tomographic reconstruction, and their interpretation. Conclusions and the scope for future work are presented in Section 9.

2. Shadowgraph Technique

Optical shadowgraphy has been used in experimental fluid mechanics research essentially for flow visualization. It employs an expanded collimated beam of light from a laser. The light beam traverses through the field of disturbance, the aqueous solution in the present application. If the disturbance is a field of varying refractive index, then the individual light rays passing through the field are refracted and bent out of their original path. This causes a spatial modulation of the light-intensity distribution. The resulting pattern is a shadow of the refractive-index field in the region of the disturbance.

2.1 Mathematical Formulation

Consider a medium with refractive index n that depends on all the three space co-ordinates, namely $n = n(x, y, z)$. We are interested in tracing the path of light rays as they pass through this medium. Starting with the knowledge of the angle and the point of incidence of the ray at the entrance plane, we would like to know the location of the exit point on the exit window and the curvature of the emergent ray [20-22].

Let the ray be incident at point $P_i(x_i, y_i, z_i)$ and the exit point be $P_e(x_e, y_e, z_e)$. According to Fermat's principle the optical path length traversed by the light beam between these two points has to be minimum. Thus

$$\delta\left(\int_{P_i}^{P_e} n(x, y, z)\, ds\right) = 0 \tag{1}$$

Parameterizing the light path by z, the above condition can be represented by two functions $x(z)$ and $y(z)$, and the equation can be rewritten as

$$\delta\left(\int_{z_i}^{z_e} n(x, y, z)\sqrt{x'^2 + y'^2 + 1}\, dz\right) = 0 \tag{2}$$

where the primes denote differentiation with respect to z. Application of the variational principle to the above equation yields two coupled Euler-Lagrange equations, that can be written in the form of the following differential equations for $x(z)$ and $y(z)$:

$$x''(z) = \frac{1}{n}(1 + x'^2 + y'^2)\left(\frac{\partial n}{\partial x} - x'\frac{\partial n}{\partial z}\right) \tag{3}$$

$$y''(z) = \frac{1}{n}(1 + x'^2 + y'^2)\left(\frac{\partial n}{\partial y} - y'\frac{\partial n}{\partial z}\right) \tag{4}$$

The four constants of integration required to solve these equations come from the boundary conditions at the entry plane of the chamber. These are the co-ordinates $x_i = x(z_i)$, $y_i = y(z_i)$ of the entry point z and the local derivatives $x'_i = x'(z_i)$, $y'_i = y'(z_i)$. The solution of the above equation yields the two orthogonal components of the deflection of the ray at the exit plane, and also its curvature on exit.

In the experiments performed, the medium has been confined between parallel boundaries and the illumination is via parallel beam perpendicular to the entry plane. The length of the growth chamber containing the fluid is L and the screen is at a distance D behind the growth chamber. The z-coordinates at entry, exit and on the screen are given by z_i, z_e and z_s, respectively. Since the incoming beam is incident normal to the entrance plane, there is no refraction at the optical window. Hence the derivatives of all the incoming light rays at the entry plane are zero; $x'_i = y'_i = 0$. The displacements $(x_s - x_i)$ and $(y_s - y_i)$ of a light ray on the screen (x_s, y_s) with respect to its entry position (x_i, y_i) are

$$x_s - x_i = (x_e - x_i) + Lx'(z_e) \tag{5}$$

$$y_s - y_i = (y_e - y_i) + Ly'(z_e) \tag{6}$$

where x_e, y_e and $x'(z_e)$, $y'(z_e)$ are given by the solutions of Eqs. (3) and (4), and the refraction at z_e, respectively.

The above formulation can be further simplified with the following assumptions:

Assumption 1. Assume that the light rays incident normally at the entrance plane undergo only infinitesimal deviations inside the inhomogeneous field but have a finite curvature on exiting the chamber. The derivatives $x'(z_i)$ and $y'(z_i)$ are therefore zero whereas $x'(z_e)$ and $y'(z_e)$ at the exit plane have finite values. The assumption is justifiable because of the smoothly varying refractive index of the KDP solution. Under this assumption, Eqs. (3) to (6) become

$$x''(z) = \frac{1}{n}\left(\frac{\partial n}{\partial x}\right) \tag{7}$$

$$y''(z) = \frac{1}{n}\left(\frac{\partial n}{\partial y}\right) \tag{8}$$

$$x_s - x_i = Lx'(z_e) \tag{9}$$

$$y_s - y_i = Ly'(z_e) \tag{10}$$

Rewriting Eqs. (9) and (10) as

$$x_s - x_i = L\int_{z_i}^{z_e} x''(z)\,dz \tag{11}$$

$$y_s - y_i = L\int_{z_i}^{z_e} y''(z)\,dz \tag{12}$$

and using Eqs. (7) and (8), Eqs. (11) and (12) become

$$x_s - x_i = L\int_{z_i}^{z_e} \frac{\partial(\log n)}{\partial x}\,dz \tag{13}$$

$$y_s - y_i = L\int_{z_i}^{z_e} \frac{\partial(\log n)}{\partial y}\,dz \tag{14}$$

Assumption 2. The assumption of the infinitesimal deviation inside the growth chamber can be extended and taken to be valid even for the region falling between the screen and the exit plane of the chamber. As a result, the coordinates of the ray on the screen can be written as

$$x_s = x_i + \delta_x(x_i, y_i) \tag{15}$$

$$y_s = y_i + \delta_y(x_i, y_i) \tag{16}$$

The deviation of the rays from their original paths in passing through the inhomogeneous medium results in the change of intensity distribution on the screen as compared to the original distribution when the disturbance is not present in the beam path. Shadowgraph method measures this change in the intensity distribution. The intensity at point (x_s, y_s) on the screen is a result of several beams moving from locations (x_i, y_i) and getting mapped on to the point (x_s, y_s) on the screen. Considering the fact that the area of the initial spread of the light beam gets deformed on passing through the refractive medium, the intensity at point (x_s, y_s) is

$$I_s(x_s, y_s) = \sum_{(x_i, y_i)} \frac{I_o(x_i, y_i)}{\left| \frac{\partial(x_s, y_s)}{\partial(x_i, y_i)} \right|} \quad (17)$$

where I_s is the intensity on the screen in the presence of the inhomogeneous refractive index field, and I_o is the original intensity distribution in the absence of the inhomogeneous field. The denominator in the above equation is the Jacobian of the mapping function of (x_i, y_i) into (x_s, y_s). It geometrically represents the ratio of the area enclosed by four adjacent rays before and after passing through the inhomogeneous medium. In the absence of the inhomogeneous field, such an area is a regular quadrilateral, which transforms to a deformed quadrilateral when imaged on to a screen in the presence of the inhomogeneous field. The summation in the above equation extends over all the ray points (x_i, y_i) which get mapped onto (x_s, y_s) on the screen and contribute to the intensity at that point.

Assumption 3. Under the assumption of infinitesimal deviations, the deflections δ_x and δ_y are small. Therefore, the Jacobian can be assumed to be linearly dependent on them by neglecting the higher powers of δ_x and δ_y, and also their product. Therefore, the Jacobian can be expressed as

$$\left| \frac{\partial(x_s, y_s)}{\partial(x_i, y_i)} \right| \approx 1 + \frac{\partial(x_s - x_i)}{\partial x} + \frac{\partial(y_s - y_i)}{\partial y} \quad (18)$$

Substituting in Eq. (17), we get

$$I_s(x_s, y_s) \left[1 + \frac{\partial(x_s - x_i)}{\partial x} + \frac{\partial(y_s - y_i)}{\partial y} \right] = \sum_{(x_i, y_i)} I_o(x_i, y_i) \quad (19)$$

Simplifying further, we get

$$\frac{I_o(x_i, y_i) - I_s(x_s, y_s)}{I_s(x_s, y_s)} = \frac{\partial(x_s - x_i)}{\partial x} + \frac{\partial(y_s - y_i)}{\partial y} \quad (20)$$

Using Eqs. (13) and (14) for $(x_s - x_i)$ and $(y_s - y_i)$, and integrating over the dimensions of the growth chamber, we get

$$\frac{I_o(x_i, y_i) - I_s(x_s, y_s)}{I_s(x_s, y_s)} = (L \times D) \left(\frac{\partial^2}{\partial x^2} + \frac{\partial^2}{\partial y^2} \right) \{ \log n(x, y) \} \quad (21)$$

Eq. (21) is the governing equation of the shadowgraph process under the set of linearizing approximations 1-3. In concise form the above equation can be rewritten as

$$\frac{I_o - I_s}{I_s} = (L \times D)\nabla^2 \{\log n(x, y)\} \qquad (22)$$

3. Numerical Solution

The governing differential equation of the shadowgraph process (Eq. (22)) relates the intensity in the shadowgraph image to the refractive index field of the inhomogeneous medium. In order to solve the equation for refractive index, a numerical procedure is followed. Here the governing equation is discretised over the physical domain of interest by a finite difference method. The resulting system of algebraic equations is solved for the image under consideration to yield a depth-averaged refractive index value for each node point of the grid. A mix of Dirichlet and Neumann boundary conditions are used for the purpose. The refractive index conditions used on the boundaries are shown in Fig. 1. A computer code has been developed for solving Eq. (22), and has been validated against analytical examples. The experimental input to the code is the matrix containing the gray value of each pixel of the shadowgraph image, and the output generated by the solver is a matrix containing the refractive index at each node point of the grid.

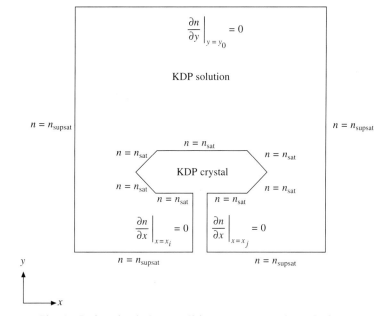

Fig. 1 Refractive index conditions used on the boundaries.

The relationship between the refractive index and the concentration of the KDP solution saturated at different temperatures is well documented in literature [23].

4. Computerized Tomography

Computerized tomography is defined as the process of producing a two-dimensional distribution of a function from its one-dimensional line-integrals obtained along a finite number of lines at known locations [14]. This could be extended to include three-dimensional reconstruction of a function from its two-dimensional projections taken along a fixed number of known directions. In the context of the present study, the line integrals of refractive index can be converted to point-wise distribution of refractive index at selected planes above the crystal.

Tomography is essentially a two-step process: first, collection of the projection data and second, reconstructing the field parameter using numerical algorithms. Reconstruction algorithms referred to as *convolution back projection* (CBP) and *algebraic reconstruction technique* (ART) are briefly reviewed as follows.

4.1 Convolution Back Projection Algorithm [24-26]

Consider reconstruction of a function $f(r, \phi)$, which in general represents physical quantities such as density, refractive index and void fraction, from its line integrals taken along a few directions. Fig. 2 shows the schematic drawing of the data collection geometry for a parallel beam CT scanner. Here, a unit circle represents the object, and the chord SD represents one data ray. The ray indices are s and θ, where s is the perpendicular distance of the ray from the object centre, and θ the angle of the source position.

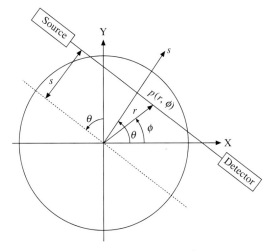

Fig. 2 Parallel beam data collection geometry.

The radon transform defines the projection data as

$$p(s, \theta) = \int_{SD} f(r, \phi)\, dz \qquad (23)$$

Implementing the *projection slice theorem*, namely the equivalence of 2D Fourier transform of the object function $f(r, \phi)$ and 1D Fourier transform of projection data $p(s, \theta)$ we get

$$\hat{p}(R, \theta) = \hat{f}(R \cos \theta, R \sin \theta) \qquad (24)$$

Taking the 2D inverse Fourier transform of the above function results in the tomographic inversion formula

$$f(r, \phi) = \int_0^\pi \int_{-\infty}^\infty \hat{p}(R, \theta) e^{i2\pi Rr \cos(\theta-\phi)} |R| dR \, d\theta \tag{25}$$

where

$$\hat{p}(R,\theta) = \int_{-\infty}^\infty \hat{p}(s, \theta) e^{-i2\pi Rs} ds \tag{26}$$

In the above equation for $f(r, \phi)$, the inner integral is divergent. Hence, for practical implementation of the above formula a filter function has to be used. The filter introduces frequency band-limitation in the reconstruction. The approximate form of $f(r, \phi)$, after introduction of the filter function is

$$\tilde{f}(r, \phi) \approx \int_0^\pi \int_{-\infty}^\infty \hat{p}(R, \theta) e^{i2\pi Rr \cos(\theta-\phi)} |R| W(r) dR \, d\theta \tag{27}$$

where

$$W(R) = 1, \quad |R| \le R_c$$
$$= 0, \quad |R| > R_c$$

Using the convolution theorem of the Fourier transforms, we get

$$\tilde{f}(r, \phi) \approx \int_0^\pi \int_{-\infty}^\infty p(s, \theta) q(s' - s) ds \, d\theta \tag{28}$$

where

$$q(s) = \int_{-\infty}^\infty |R| W(R) e^{i2\pi Rs} dR \tag{29}$$

with $p(s, \theta)$, the projection data; s', the s value of the data ray passing through the point (r, ϕ) being reconstructed; θ, the source position; R, R_c, the Fourier frequency and the Fourier cut-off frequency, respectively; $q(s)$, the convolving function; and $W(R)$, the filter function.

The inner integral in the final equation is one-dimensional convolution, and the outer integral, corresponding to the averaging operation (over θ), is termed as the back projection and hence the name *convolution back projection* for this reconstruction algorithm.

4.2 Algebraic Reconstruction Technique

The algebraic reconstruction technique [27, 28] is fundamentally different from the transform technique in the sense that the discretization of the problem is carried out at the very beginning. In the transform methodology of Section 4.1, the continuous problem is analytically treated until the very end, when the final formulas are discretized for computational implementation. Fig. 3 shows the discretization of the region of interest.

Collection of projection data results in a matrix equation

$$\{\Phi_i\} = [w_{ij}] \{f_j\} \tag{30}$$

where $j = 1...N$, $i = 1...M$ $(> N)$, with f representing the basic unknowns of the grid, Φ, the projection data, and $[w_{ij}]$, the weight factor. The above problem reduces to the problem of finding a generalized inverse of the matrix $[w_{ij}]$. The steps involved in the algebraic algorithm are as follows

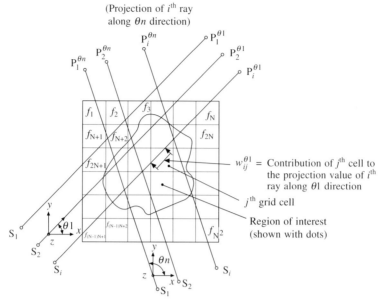

Fig. 3 Discretization of the region of interest.

1. Initial assumption of the field to be reconstructed over a grid.
2. Calculation of the projection data with the initial guess.
3. Compute the correction factor.
4. Distribute correction factor suitably among all the cells.
5. Test for convergence.

5. Apparatus and Instrumentation

5.1 Crystal Growth Apparatus

Two different crystal growth chambers are designed and fabricated for performing the experiments. The first growth chamber has relatively simple design with two concentric beakers, one inside the other. The crystal growth occurs inside the inner beaker whereas the cavity between the two beakers is filled with thermostated water to keep the KDP solution at the desired temperature. In order to facilitate unhindered passage of the laser beam through the growth chamber, both beakers are provided with beam ports covered by optical windows. Specific locations are provided on the growth chamber for inserting a temperature sensor and a KDP seed crystal mounted on a platform. The experiments reported in the present paper were conducted using this growth chamber. Since only one through port for the laser beam is provided on the growth chamber, the tomographic projections are recorded by turning the crystal in discrete steps. A second growth chamber with several novel design features is fabricated in order to record the tomographic projections without disturbing (turning) the crystal during growth. In addition to the feature of not disturbing the crystal during experimentation, the cell has design features, incorporated keeping in view the extremely high level of instrumental precision required in Mach-Zehnder interferometric tomography experiments. The growth chamber has double walled geometry with the inner cell acting as the growth cell and the cavity between the inner and the outer cell filled by thermostated water to maintain the temperature of the KDP solution at the desired value. The inner cell is an octagonal shaped glass vessel having one beam port on each side of the octagon which are covered by quartz optical windows for the passage of the laser beam. The octagonal cell is placed inside a cylindrical glass vessel that has eight beam ports. The passage of the laser beam from the outer cell to the inner cell is isolated from the intervening thermostated water by aluminium tunnels connecting the two cells. The parallelism of the four pairs of windows is ensured using auto-collimator during fabrication. The residual wedge, if any, between the pair of windows on the inner cell is compensated by using wedge-compensators, one each for the eight beam ports. Wedge compensators are particularly necessary for the interferometric tomography experiments wherein the alignment of the interferometer has to be maintained either in the wedge setting mode or the infinite setting mode for the entire duration of the experiment. The growth chamber is placed on a high-precision wobble-free motorized rotation assembly for turning the growth chamber and collecting the projection data. The eight-window growth chamber with wedge-compensators mounted on each beam port and the chamber placed on the rotation assembly is shown in Fig. 4.

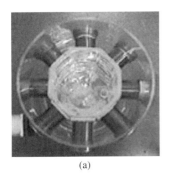
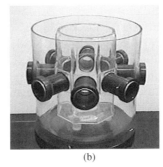
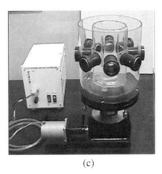

(a) (b) (c)

Fig. 4 Eight-window crystal growth chamber. (a) Top view of the cell, (b) side view of the chamber and (c) chamber placed on a motorized rotation assembly.

5.2 Temperature Controller
The stability of the solution temperature is crucial to conduct experiments with repeatability, and to avoid the problem of spurious nucleation. The crystal growth literature points towards fluctuating temperatures as the possible cause of the inclusion formation. Thus, stability of the temperature of the solution is a necessity for growing good quality crystals. The design of the growth cell described above was mainly driven by this consideration. In addition to the specific design, a PID based temperature controlling instrumentation [*Eurotherm*] was used. The signal from the temperature controller is dependent on the output of a comparator circuitry, which continuously compares the set-point temperature with the process temperature measured by a resistance thermometer. Solution temperature was maintained to within $\pm 0.01°C$ in the present experiments.

5.3 Optical Set-up and Data Acquisition Instrumentation
Fig. 5 shows the schematic layout of the optical and the image acquisition instrumentation. A He-Ne laser of 5 mW power is expanded and collimated to a diameter of 30 mm by a beam expander. The collimated beam passes through the crystal growth chamber in which the KDP crystal is grown. The beam emerging at the exit window of the growth cell falls on a screen

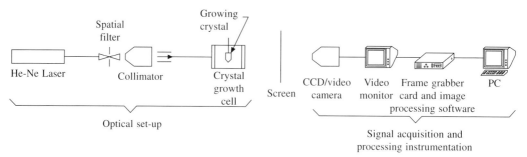

Fig. 5 Optical and image acquisition instrumentation.

resulting in the shadowgraph image. The images are continuously recorded using a video camera (*Sony Handycam* 360X) interfaced to a computer through a 1024 × 1024, 25 Hz frame grabber card.

6. Experimental Procedure

In order to grow a KDP crystal from its aqueous solution, a saturated solution of KDP chemical was prepared using de-ionized water having an electrical resistivity of 18 MΩ-cm and chemicals having purity of 99.9 %. The saturation temperature of the solution was measured as 56°C. The solution was filtered using special membrane filter, 0.02 μm pore size, and then over-heated by 20°C above the saturation temperature for 24 h to dissolve any possible clusters. This elaborate procedure reduces the problem of spurious nucleation. In order to initiate the growth, a full morphology KDP seed crystal of size 2 × 2 × 2 mm^3 was mounted on the tip of a 1.5 mm diameter glass rod. The glass rod with seed mounted on top of it was fixed in an upright position on a horizontal platform made from Plexiglas. The whole assembly was then immersed inside the growth chamber containing the saturated solution of KDP. The growth started at 56°C and the temperature of the solution was reduced at the rate of 0.02°C/h at appropriate stages of growth. The growth run lasted for 195 h before it was terminated at 50°C. During the entire duration of growth, no spurious nucleation inside the crystallizer was observed, indicating that the growth was performed in a highly controlled manner and the saturation temperature was accurately determined.

7. Shadowgraph Experiments

In the presence of the immersed crystal, the KDP solution is a field of varying refractive index, and light rays passing through it are refracted out of their original path. This causes a spatial modulation of the intensity distribution with respect to the original intensity on the screen, and we get a shadowgraph pattern on the screen. The development of buoyancy driven convection was recorded at regular intervals in the form of shadowgraph images for the entire duration of the experiment.

8. Results and Discussion

8.1 Visualization of the Buoyancy Driven Convective Activity

Fig. 6 (a-f) show the shadowgraph images of the convection process taken at regular intervals during the 195 hours of growth. The plume activity gradually builds up as the crystal dimension increases. Initially the plumes due to buoyancy driven convection are weak in intensity, but stable and laminar as shown in Fig. 6 (a). After about 50 h of growth the crystal dimension increases and the plumes become strong. We observe that the plumes are seen to rise only from the left and the right sides of the crystal. This is because the depleted solution moves along the crystal surface, and at the edges, rises in the form of two streaks of lower density solution. The laminar behaviour of plumes is maintained at this stage as well as shown in Fig. 6 (b, c). After about 100 h of growth, plumes are seen to emerge from several spots on the crystal top surface. This indicates that the growth rate has increased and there are several locations on the top surface where there is accumulation of the solution depleted of solute that are sources of plumes. At this stage the plume behaviour breaks from its laminar nature and becomes irregular as shown in Fig. 6 (d). After about 190 hours of growth, the buoyancy driven convective activity is chaotic, as seen in the plume behaviour (Fig. 6 (e-f)). This is the final state of buoyancy driven flow observed in our experiments.

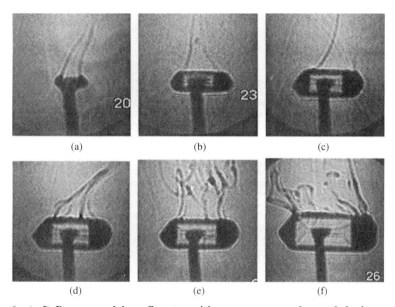

Fig. 6 (a-f) Buoyancy driven flow transition sequence as observed during growth. The activity makes transition from being laminar in the beginning to chaotic at the end, intermediated by irregular flow.

8.2 Projection Data

After 160 hours of growth the buoyancy driven convective activity is strong, as plumes of broad width at the base are rising continuously from the top surface of the growing crystal. At this stage, 19 shadowgraph images are collected by turning the crystal in small steps of 10° between 0 and 180°. The images shown in Fig. 7 (a-h) are shadowgraphs recorded at 0, 20, 40, 60, 80, 140, 160 and 180°, respectively. The set of 19 projection images constitute the experimental projection data for performing tomographic reconstruction of the concentration field above the crystal surface.

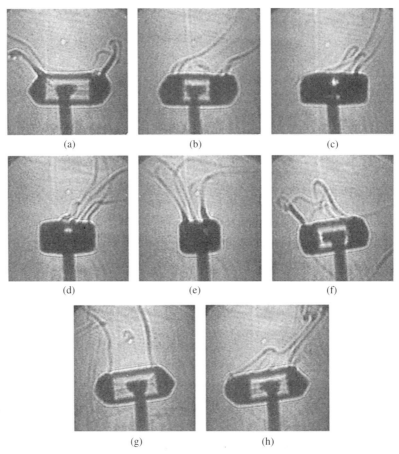

Fig. 7 (a-h) Shadowgraph images recorded along 0, 20, 40, 60, 80, 140, 160 and 180° angles, respectively, showing the buoyancy driven convection plumes above the top surface of a growing KDP crystal. Differences in the shadowgraph images among the various view angles show that the convective field is three-dimensional.

8.3 Two-dimensional Refractive Index Field

Shadowgraph images have been analyzed numerically as per the procedure described in Section 3 to obtain the refractive index field. Fig. 8 (a-h) are the refractive index contour maps corresponding to the shadowgraph images of Fig. 7 (a-h), respectively. A row in the refractive index matrix represents a particular plane through the growth chamber. The entries in the row are line integral values of the refractive index field averaged along the viewing direction of the laser beam.

8.4 Reconstruction Using Convolution Back Projection (CBP) Algorithm

Implementation of the CBP algorithm demands continuous projection data between 0 and 180°. The experimental set-up used in the present work allowed only 19 projection images between 0 and 180°. Therefore, in order to implement CBP, an interpolation scheme is needed for increasing the size of the projection data set.

In tomography, the projection data required for reconstruction should be available for the full width of the flow field from each view angle rather than the partial width. The experimental apparatus used in the present work was such that only 20% of the flow field could be covered in each view angle, owing to the size of the laser beam (30 mm) being less than the diameter of the process chamber (150 mm). It resulted in a partial projection data set. This limitation was overcome by following a suitable extrapolation scheme to obtain projection values for points lying outside the volume included by the source-detector configuration. The extrapolation procedure adopted in the present work assumes that the concentration field is spatially uniform outside the central core of 20%. This is a justifiable assumption as the solution concentration is practically homogeneous away from the crystal-solution interface. However, the variations in the chord length of the rays as they pass through different regions of the growth chamber have to be taken into account. This procedure is not expected to alter the information content of the original projection data obtained through experiments. The implementation of the interpolation step

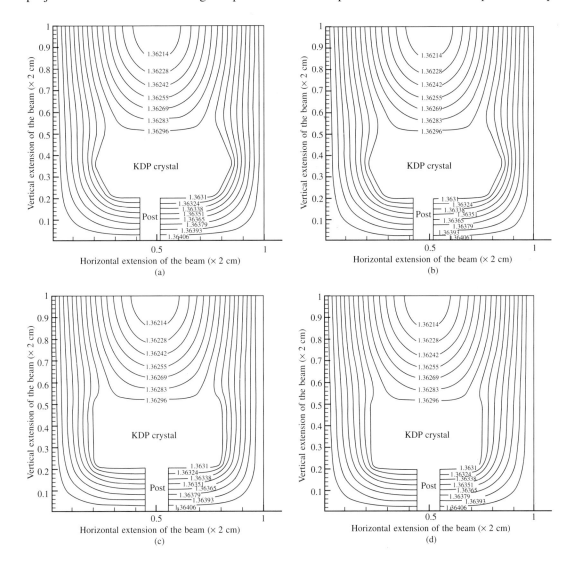

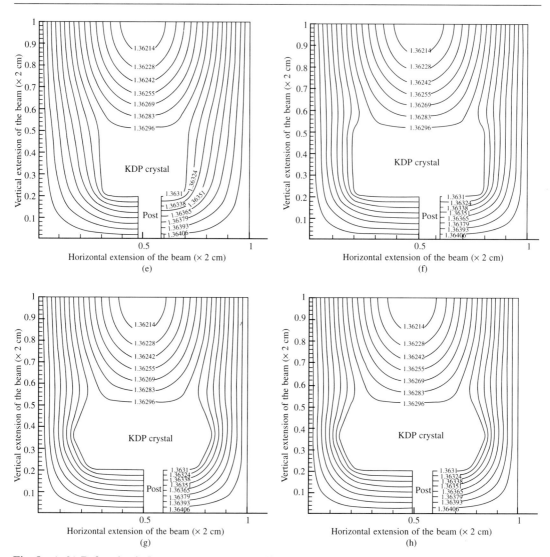

Fig. 8 (a-h) Refractive index contours corresponding to the shadowgraph images shown in Fig. 7 (a-h), respectively. They are obtained by solving the Poisson equation governing the shadowgraph process.

followed by extrapolation of the experimental information, results in *complete* projection data, an essential requirement for the CBP algorithm.

The projection data for each view angle should be consistent in terms of the average solution concentration over a given plane of the fluid layer. In other words, the average concentration should be constant for a given plane irrespective of the view angle. This mass-balance consistency check was enforced on the data set before applying reconstruction algorithms. CBP based reconstruction of the concentration field over a plane at a height of 0.15 (normalized) above the crystal surface is shown in Fig. 9 (a).

8.5 Reconstruction Using Algebraic Reconstruction Technique (ART)

The discussion on the need for a consistent data set is applicable for the ART family of algorithms as well. However, these algorithms can work with partial data. The solution obtained by this approach is expected to capture the global trends in the numerical solution, with finer features appearing in the reconstruction, when more complete projections become available.

Implementation of the ART family of algorithms requires an initial guess for the refractive index field to start the iterations. Convergence is known to be strongly sensitive to the starting guess. One of the purposes of using CBP in the present work, in spite of the recommendation against its use in situations involving incomplete and partial projection data, was to use this solution as the starting guess for the reconstruction based on ART. The ART family of algorithms have the advantage that they facilitate a crosscheck on the correctness of the converged solution. The procedure for the crosschecks has been discussed in detail by Mishra et al [18-19]. ART based reconstruction of the concentration field over a plane at a height of 0.15 (normalized) above the crystal surface is shown in Fig. 9 (b).

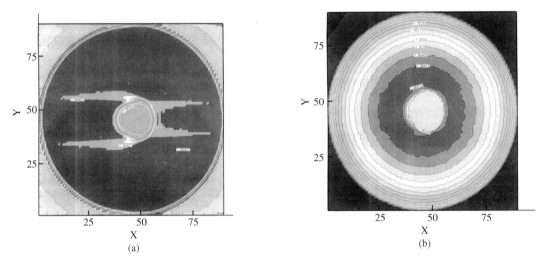

Fig. 9 Reconstructed concentration field over a plane at a height of 0.15 (normalized) above the crystal surface using: (a) CBP algorithm and (b) ART algorithm, respectively.

8.6 Growth Rate and the Quality of the Crystal

The growth experiment was performed for 195 h during which the crystal grew in dimensions from 2 mm^3 to $17 \times 9 \times 7$ mm^3. This corresponds to a growth rate along c-axis of approximately 2 mm per day, a value comparable to that reported in the literature [23] for conventional growth under free convection. Fig. 10 (a, b) show the crystal perched on top of a capillary fixed on a platform, and after removal from the platform, respectively. The crystal was clear of any visible defects such as inclusions, bubbles and streaks except those present in the starting seed crystal. The percentage transmission in the visible region was measured using spectrophotometer and was found to be over 80% (with correction for Fresnel losses) revealing the good optical quality of the grown crystal. Fig. 10 (c) shows the high transparency of the crystal grown in the present experiment.

Fig. 10 (a-c) Crystal on the platform immediately after removal from the growth chamber indicating size of the crystal and transparency of the grown crystal.

9. Conclusions

We have visualized buoyancy driven convection during the growth of a KDP crystal from its aqueous solution using the shadowgraph technique. In order to perform tomographic reconstruction, special crystal growth chambers with novel design features have been fabricated. Shadowgraph projections are recorded, and analyzed numerically to obtain 2D refractive index fields. The applicability of Fourier transform (CBP) and series expansion (ART) based algorithms for 3D reconstruction of the concentration field over the crystal surface inside the growth chamber are presented. Experiments show that the convective activity gradually undergoes transition from a laminar to a chaotic state, intermediated by irregular behaviour at different stages of the growth. The growth rate was comparable to the value reported in literature, and the optical quality of the grown crystal as characterized through transmission was found to be good.

Acknowledgement

The authors are grateful to Prof. Prabhat Munshi for introducing them to the field of computerized tomography and for useful discussions during the course of the present work.

References

1. F. Rosenberger and G. Muller, Influence of convection on the growth of crystals from solution, *J. Cryst. Growth*, **65** (1983) 91.
2. W.R. Wilcox, Transport phenomena in crystal growth from solution, *Prog. Cryst. Growth and Charact.*, **26** (1993) 153.
3. F. Rosenberger, Boundary layers in crystal growth – facts and fancy, *Prog. Cryst. Growth and Charact.*, **26** (1993) 87.
4. Y. Zhou and J.J. Derby, Three-dimensional computations of solution hydrodynamics during the growth of potassium dihydrogen phosphate. Part-I: Spin up and steady rotation, *J. Cryst. Growth*, **180** (1997) 497.
5. A. Yeckel, Y. Zhou, M. Dennis and J.J. Derby, Three-dimensional computations of solution hydrodynamics during the growth of potassium dihydrogen phosphate. Part-II: Spin down, *J. Cryst. Growth*, **191** (1998) 206.
6. B. Vartak, Yong-Il Kwon, A. Yeckel and J.J. Derby, An analysis of flow and mass transfer during the solution growth of potassium titanyl phosphate, *J. Cryst. Growth*, **210** (2000) 704.
7. H.F. Robey and D. Maynes, Numerical simulation of the hydrodynamics and mass transfer in the large scale, rapid growth of KDP crystals. Part 1: Computations of the transient, three-dimensional flow field, *J. Cryst. Growth*, **222** (2001) 263.

8. H.F. Robey, Numerical simulation of the hydrodynamics and mass transfer in the large scale, rapid growth of KDP crystals. Part 2: Computation of the mass transfer, *J. Cryst. Growth,* **259** (2003) 388.
9. N. Zaitseva and L. Carman, Rapid growth of KDP-type crystals, *Prog. Cryst. Growth and Charact.,* **43** (2001) 1.
10. W. Merzkirch, *Flow Visualization* (Academic Press, London, 1987).
11. R. Goldstein (Ed.), *Fluid Mechanics Measurements* (Taylor & Francis, U.K., 1996).
12. G.S. Settles, *Schlieren and Shadowgraph Techniques* (Springer, Berlin, 2001).
13. F. Mayinger (Ed.), *Optical Measurements* (Springer-Verlag, Berlin, 1994).
14. G.T. Herman, *Image reconstruction from projections: The fundamentals of computerised tomography* (Academic Press, New York, 1980).
15. Sunil Verma, K. Muralidhar and V.K. Wadhawan, Flow visualization and modeling of convection during growth of KDP crystals, *Ferroelectrics,* **323** (2005) 25–37.
16. P. Munshi, Error analysis of tomographic filters, *NDT & E International,* **25** (1992) 191.
17. P. Munshi, M. Maisl and H. Reiter, Experimental aspects of the approximate error formula for computerized tomography, *Materials Evaluation,* **55(2)** (1997) 188.
18. Debasish Mishra, Experimental study of Rayleigh-Benard convection using interferometric tomography, *Ph.D. Thesis,* IIT Kanpur, India, 1998.
19. Debasish Mishra, K. Muralidhar and P. Munshi, Interferometric study of Rayleigh-Benard convection using tomography with limited projection data, *Expt. Heat Transfer,* **12** (1999) 117.
20. M. Born and E. Wolf, *Principles of Optics* (Pergamon Press, U.K., 1980).
21. W. Schopf, J.C. Patterson, and A.M.H. Brooker, Evaluation of the shadowgraph method for the convective flow in a side heated cavity, *Exp. Fluids,* **21** (1996) 331.
22. B.L. Winkler and P. Kolodner, Measurements of the concentration field in non-linear travelling-wave convection, *J. Fluid Mech,* **240** (1992) 31.
23. L.N. Rashkovich, KDP-Family Single Crystals (Adam-Hilger, New York, 1991).
24. G.N. Ramachandran and A.V. Lakshminarayanan, Three dimensional reconstruction from radiographs and electron micrographs: Application of convolution instead of Fourier transforms, *Proc. Natl. Sci. Acad., USA,* **68** (1970) 2236.
25. R.M. Lewitt, Reconstruction algorithm – transform methods, *Proc. of IEEE,* **71** (1983) 390.
26. M.F. Manzoor, Pankaj Yadav, K. Muralidhar and P. Munshi, Image reconstruction of simulated specimens using convolution back projection, *Defence Science Journal,* **51** (2001) 175.
27. Y. Censor, Finite series-expansion reconstruction methods, *Proc. of IEEE,* **71** (1983) 409.
28. K. Muralidhar, Temperature field measurement in buoyancy-driven flows using interferometric tomography, *Annual Reviews of Heat Transfer,* **12** (2001) 265.

Computerized Tomography for Scientists and Engineers
Edited by P. Munshi
Anamaya Publishers, New Delhi, India

15. Digital Radiography for Non-destructive Testing

Debasish Mishra, Rajashekar Venkatachalam and V. Manoharan

Industrial Imaging and Modeling Laboratory, John F. Welch Technology Center,
General Electric Company, Bangalore-560 066, India

Abstract

Digital radiography is rapidly taking over film radiography for industrial applications. Digital radiography is in the process of becoming the industry standard. It is already accepted as a preferred x-ray imaging choice for medical diagnostics. There are different forms of digital radiography imaging systems available commercially and in the absence of a clear understanding of the systems it is often a challenge to select the right system for an application. Productivity benefit is the major advantage of digital system over traditional film systems. Combination of micro-focal x-ray tube with digital recording device has the capability to match with film resolution. Short exposure and higher contrast-to-noise ratio gives advantage for fast and quality inspection.

The subject of digital radiography is not new. However, a detailed description of various digital recording devices and image quality measurement specifically for industrial applications is not available in one place for a standard reference. The objective of this article is to introduce the concepts of various digital recording devices to the user and describe various image quality metrics that can be used to compare the performance of the systems. The chapter also discusses artifacts seen in various digital systems, noise performance and a guideline that can be used for selecting the right kind of imaging device for a specific application.

1. Introduction

Since the discovery of x-rays in 1895, x-ray radiography has been one of the widely accepted non-destructive testing (NDT) methods in industry. X-ray radiography is an imaging technique, which produces internal evaluation of objects. Film radiography, in particular where a film is used to record the penetrating x-ray radiation, is the backbone of industrial NDT applications.

In radiography, images are formed due to differential attenuation of x-rays penetrating through the object under investigation. More the differential attenuation, better is the quality of the acquired image. The radiographer usually selects the x-ray energy, x-ray photon flux and time of exposure such that they maximize the image quality. Noise in the image formation process is inherent and there are many sources responsible for it. The emission of x-rays and absorption of x-rays in materials is a statistical process. Poisson distribution effect is seen on the recorded x-ray images since we integrate the photon flux on the detectors for a finite time. This photon arrival variation causes random variation in the images, which are Poisson in nature, and is a noise source for the image. Similarly, other factors related to the recording device can also cause noise formation in the images; these factors can be electronic noise, chemical variation, film non-uniformity etc. The available contrast, which is a measure of the signal level in an image has to be several times more than the noise level to increase the visibility of the features in an image and hence related to the detectability of features in x-ray images. Detectability is, thus, related to a parameter usually referred to as contrast-to-noise ratio (CNR). Higher CNR is required to make an imaging system adequate for industrial inspection.

The ability to resolve smaller features with sufficient CNR is another important criterion for inspections. Small size defects (usually a few microns) are the major bottlenecks for many manufacturing processes. Similarly, cracks initiation (small features) in many industrial components if detected at an early stage can save lives and avoid disaster. While large CNR for smaller features are important, higher throughput is also a major constraint that needs to be satisfied. This demands image acquisition during inspections to be faster. Image acquisition can be made faster with recording device having enough sensitivity at lower dose so that exposure time can be cut down. Further throughput can be increased several folds by having higher latitude coverage in the detecting media. Higher latitude mainly refers to covering larger variation in depths of material being inspected in one image. If this condition is not met, a component needs to be imaged with different settings to cover the thickness variations.

Many regulatory organizations require images acquired for inspections to be stored for some time. They compare the health of an industrial component over a period of time to check its reliability for future usage. Image archival is thus an important area for x-ray-based NDT. Lower cost of the system, lower operational cost and automation of the process has always been the voice of the customer. Further, more adaptable x-ray imaging system with generic and easy-to-use recording device is a constant demand from the users. As we advance through parallel growth in various technology areas, the use of x-ray imaging also has evolved from a traditional projection imaging to various new dimensions. Volumetric computed tomography, energy discriminative radiography, phase contrast imaging and reverse geometry imaging, are some of the examples of growth in the area of x-ray imaging. We need imaging systems, which are easily adaptable to such new applications.

Finally, the images that are formed in the recording device should be easy for interpretation by an operator. Operator's interpretation can have subjectivity when the inspections are manual. Manual inspections can also lead to errors due to operator fatigue. Hence, there is a need for automating the inspection process, especially the image interpretations.

Excellent image quality (high CNR), higher resolution capability (smaller features with enough CNR), higher throughput (less exposure time and better latitude), lower cost, easy image archivals, adaptable for new and advanced applications and automation for image interpretations are some of the major drives for x-ray imaging technology. Over a period of time the recording devices used for radiography inspections have emerged and are continuously striving to satisfy the end users requirements. Digital radiography (DR) is a result of this revolution.

From traditional film-based x-ray imaging we have today emerged into flat panel-based digital radiography with real time imaging capabilities. Image intensifiers, storage phosphors, charged-coupled device (CCD) imaging are some of the intermediate developments that we have seen. This article discusses various digital options that are available today. Section 3 explores the scintillator materials used in various digital imaging devices. With many possibilities of scintillator materials and imaging techniques that can provide digital images, performance study and selection of imaging device for a particular application is important. This can be achieved through an image quality metric. We have discussed image quality metric suitable for radiography and also presented selection criteria of detectors for different applications.

2. History of Development

X-ray image receptors evolved from film recording to the real time digital imaging in the last century. Radiology (x-ray imaging) began as a medical specialty in the first decade of the 1900's after the discovery of x-rays by Professor Roentgen. The development of radiology grew at a good pace until World War II. Slowly, the industrial inspections using x-rays also gained popularity. The demand for more efficient and fast image receptors was growing from both medical and industrial side. For the first fifty years of radiology in medicines and in industrial inspections, the primary examination involved creating an image by focusing x-rays through the body part or an industrial object directly onto a film inside a special cassette. In the earliest days, a head x-ray could require up to 11 minutes of exposure time. Now, modern x-ray images are made in milliseconds and the x-ray dose currently used is as little as 2% of what was used for 11-minutes head examination 50 years ago. Industrial inspections using films in those days required even longer exposure times.

The next development in the x-ray detectors area came with the use of fluorescent screens and special glasses so that x-ray images could be seen in real time. This technique required humans to stare directly into the x-ray beam, resulting in unwanted amount of exposure to radiation. In 1946, George Hoenander developed the film cassette changer, which allowed a series of cassettes to be exposed at a movie frame rate of 1.5 cassettes per sec. By 1953, this technique was improved to allow frame rates up to 6 frames per sec by using a special *cut film changer*.

A major development along the way for medicine was the application of pharmaceutical contrast medium to help visualize organs and blood vessels with more clarity and image contrast. These contrast media agents (liquids also referred to as *dye*) were first administered orally or via vascular injection between 1906 and 1912 and allowed doctors to see the blood vessels, digestive and gastro-intestinal systems, bile ducts and gall bladder for the first time.

In 1955, the x-ray image intensifier (also called XII) was developed and allowed the pick-up and display of the x-ray movie using a television (TV) camera and monitor. By the 1960's, the fluorescent system (which had become quite complex with mirror optic systems to minimize patient and radiologist dose) was largely replaced by the image intensifier/TV combination. Together with the cut-film changer, the image intensifier opened the way for a new radiologic sub-specialty known as angiography to blossom and allowed the routine imaging of blood vessels and the heart. Meanwhile XII also became popular in industrial inspections and even today are in use. Real-time x-ray imaging capability is a major advantage for productivity benefits where hundreds of industrial components are manufactured and needs to be inspected immediately on the production line.

Computed radiography (CR) using the storage phosphor was the next imaging technology that hit the market for x-ray inspections. The fundamental innovation in the development of CR was done by Kodak who conceived the storage of an x-ray image in a phosphor screen. It required significant technical steps and conceptualization of the application by Fuji to produce the first medical x-ray images. Fuji, the main developer of CR in the eighties, used $BaFBr:Eu^{2+}$ phosphor and a cassette-based approach. CR is in use more in industrial inspections than in medical imaging. New research efforts have made CR technology a viable and easy-to-use x-ray imaging technique.

Flat panel detectors using direct or indirect conversion of x-rays to electrical signals are the latest available technology in DR. This is possible using the amorphous silicon (a-Si) photo-diode technology with cesium iodide as the scintillator material or the selenium, which converts directly x-rays to electrons. Lower dose and higher signal-to-noise ratio and fast imaging are some of the major advantages associated with this technology. Direct availability of digital images has opened up a whole lot of different new applications such as quantitative radiography, volumetric computed tomography and automatic defect recognition in the industrial world. Other direct digital options are CCD and CMOS (complementary metal oxide semiconductor) based x-ray detectors.

Digital radiography is in the process of becoming the industry standard. It is already accepted as a preferred x-ray imaging choice for medical diagnostics. Although the resolution capability of these DR systems is not directly comparable with that of film system it is possible to design a system with suitable x-ray source like a micro-focal x-ray source to get close to that of a film system from resolution stand-point. Image quality, real-time imaging, image archival and automation of procedures are major advantages of the present day DR system.

3. Scintillator for Digital Radiography

As pointed out in the earlier section, film radiography has been in practice for several years. The major disadvantages of film radiography include difficulties in archiving, inapplicability of post-image processing enhancements and decreased productivity due to the time involved in processing them. The majority of these can be overcome, if the x-ray image is digitized. Scintillators are specific kinds of materials that can convert x-ray photons to visible light photons which then can be converted to a digital signal using technologies such as CCD or CMOS or a-Si (amorphous silicon) flat panels and an analog to digital converter. This will facilitate real time acquisition of images without the need for offline processing as in films. Since there are various stages of conversion to record the digital image, it is very important to ensure that minimum information is lost during conversion. The performance of the system needs be evaluated based on the useful information one can gather from the digital image. The various metrics used to judge the performance of a system from the digital image is explained in Section 5. This section is subdivided into four sections. The first sub-section discusses the properties of an ideal scintillator and the state-of-the-art of the available scintillators. The second sub-section explains how a scintillator works. The third sub-section presents integration of CsI (cesium iodide) scintillators with CMOS, CCD or a-Si arrays. The fourth sub-section describes the electronic read-out process, which actually generates the digital image.

3.1 Scintillator Materials

While the search for the ideal scintillator continues, in practice we must settle on a scintillator based on trade-offs between properties that are required for the targeted application and the economics involved. With the availability of more efficient and cost effective scintillator material, the x-ray detector technology has also evolved.

3.1.1 Properties of an Ideal Scintillator

As pointed by David et al [1], an ideal scintillator should satisfy the following properties:

- High stopping power for x-rays
- High x-ray photon to light photon conversion. This is measured in photons/MeV
- Emission spectrum of the scintillator should match with the peak efficiency wavelengths of the light detector employed
- The index of refraction should match between the interface and the light sensor
- Minimal decay time with no after glow
- Stable output to long exposure to radiation
- Independence of temperature on light output
- Stable mechanical and chemical properties

These properties allow for a high efficiency, stable and robust scintillator material with ideal imaging requirements (like high Modulation Transfer Function), environment friendliness and stability in their performance.

3.1.2 Performance of Commercially Available Scintillators

The performance of commercially available scintillators is compared based on the factors given in Table 1. The light yield is a measure of the capability of the scintillating material to convert radiation energy to fluorescent radiation in or near the visible region. The light yield is an important factor for the energy resolution; the more is the energy transferred to the electronics, the lower is the statistical variations in the detected signal. CsI has a wavelength of maximum emission, which matches with silicon photodiodes (or any other semiconductor material used). When atoms in a scintillator are excited they decay spontaneously. The time between the arrival of radiation in the detector and the formation of signal, in this case the time between excitation and the first photons emitted, is the response time. The response time determines the time resolution, a parameter of great significance for real-time detector. Scintillation detectors generally have fast response times. A short decay time increases the maximum count rate and is, therefore, an important property for detection of high-flux radiation.

Table 1. Properties of selected scintillators [1]

Material	Light output (photons/MeV)	Wavelength of max. emission (nm)	Decay constant (nsec)	Density (g/cc)	Index of refraction	Moisture sensitivity	% of signal after glow (msec)
NaI(Tl)	38,000	415	230	3.67	1.85	High	0.3-5/6 msec
BGO	9,000	480	300	7.13	2.15	None	0.005/3 msec
CsI(Tl)	59,000	560	1000	4.51	1.84	Slight	0.5-5/6 msec
CdWO$_4$	15,000	480	1100/14500	8	2.2	None	0.1/3 msec
CaF$_2$(Eu)	19,000	435	940	3.19	1.44	None	< 0.3/6 msec
GOS*		510	3000	7.34	2.2	None	< 0.1/3 msec
LSO**	30,000	420	40	7.4	1.82	None	
Plastics	~10,000	420	2-17	1.03	1.58	None	< 0.1/3 msec

*Gd$_2$O$_2$S with dopants: properties vary with dopant types and levels
**Lu$_2$(SiO$_4$)O:Ce

Density of the scintillator material plays a very important role in deciding the stopping power for x-rays. For high-energy x-ray applications, scintillators with high effective atomic number need to be selected. Scintillator materials typically need to have a high index of refraction, so that it is relatively easy to shape a piece of scintillator and much of the light produced inside is reflected off its surfaces and back into the material, and is eventually funneled into a detector. Among the various scintillator materials given in Table 1, CsI(Tl) seems to be a suitable candidate because of its high conversion efficiency, matching of the emission spectrum to silicon photodiodes and lower cost. It has a high after glow of 0.5-5% per msec, which restricts its use to lower x-ray energies. After glow is defined as the fraction of scintillation light still present for a certain time after the x-ray excitation stops. CsI(Tl)'s slight sensitivity to moisture can be controlled with some protective measures. Various modifications to the fabrication of CsI(Tl) structures have been proposed for operating at a wide range of x-ray energies. A wide range of mechanical and chemical variations are the key to success for designing a robust industrial x-ray detector.

3.1.3 *Schematic of a Scintillator Assembly*

A schematic of a scintillator assembly in a digital x-ray detector is shown in Fig. 1. The top most layer is the x-ray converter which can either be a scintillator or a photoconductor. Scintillator uses indirect conversion technique and photoconductors use direct conversion technique. Amorphous silicon active matrix is the charge collector. The signal generated is boosted using an amplifier and then converted to a digital signal using an analog to digital converter.

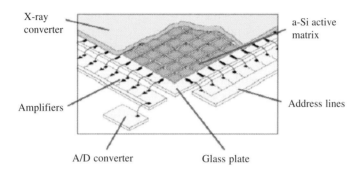

Fig. 1 **General architecture of a scintillator assembly.**

3.2 Most Commonly Used Scintillator Materials

This section discusses some of the common scintillator materials that are available with emphasis on phosphors and CsI scintillators. Gadolinium oxysulfide doped with terbium (Gd_2O_2S: Tb) is one of the efficient scintillators available in terms of light output per incident x-ray photon. It is an effective absorber to x-rays because of its high atomic number and density. However, its main limitation is that it is manufactured as a compressed powder, which means that any light generated by x-rays is rapidly scattered and diffused before it is intercepted by the detector. Cadmium telluride and related compounds such as cadmium zinc telluride are being explored as detector materials for future 2D area detectors. The advantage of using these materials is a higher absorption coefficient resulting in a higher sensitivity compared to silicon devices.

Phosphors: Phosphors have the property to emit light, when an x-ray is absorbed. This light once formed can scatter multiple times off the phosphor grains before it escapes the screen. This process is shown schematically in Fig. 2. The scattering causes image blur or resolution loss. Thicker phosphor screens absorb x-rays better but cause more blurring due to excessive light spreading.

Storage phosphors: When x-ray is incident on storage phosphors, phosphorescence occurs and some of the charge also gets trapped in the color centers of the crystalline structure. In phosphorescence, light emitted by an atom or molecule persists after the exciting source is removed. Color center refers to a lattice defect

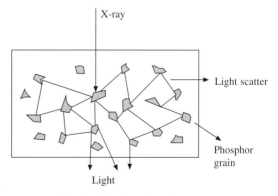

Fig. 2 **Powder phosphor screen.**

in a crystalline solid consisting of a vacant negative ion site and an electron bound to the site. Such defects will absorb light and make certain normally transparent crystals appear colored. This energy is stored as a *latent image* and can be read out later. During this read-out process the storage phosphor is stimulated by a scanning red laser and electrons return to a more stable energy level and emit blue light (photo stimulated luminescence). A light guide and a photomultiplier tube collect this light, which is proportional to the absorbed x-ray intensity. The output of this photomultiplier is amplified and digitized. The advantages of the storage phosphor in comparison to films are its wide latitude and linear response to radiation. Computed radiography (CR) relies on this storage phosphor technology, which is discussed in detail in Section 4. Despite the development of other approaches to digital radiography, CR continues to be an important means of generating digital radiographic images because of its portability, affordability, ease of use, and its ability to be easily integrated into existing x-ray facilities.

CsI Scintillators: CsI when excited by x-rays return quickly to the lower energy state and emit visible light during the de-excitation process. Radiation can be detected by measuring this light output. This is usually performed using a TFT (Thin Film Transistor) readout circuit. The lateral diffusion of the light will reduce the sharpness and spatial resolution of the image [2]. In order to overcome this, some indirect conversion detectors use structured scintillators consisting of CsI crystals that are grown perpendicular to the detector surface as shown in Fig. 3. CsI when doped with thallium (Tl) can increase the light output of the scintillator. The CsI(Tl) micro-columnar structures are high-density fibers of CsI(Tl) scintillators resulting from growing the structure on a specially designed substrate. This scintillator material is grown in preferential microstructured columns, which reduces the width of the point response function, resulting in superior spatial resolution [3].

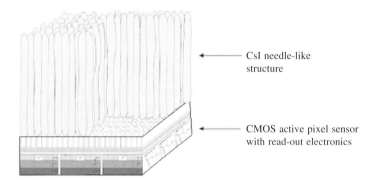

Fig. 3 Needle-like structure of thalium doped CsI structure.

3.3 Integration of Image Sensors to Scintillators

This section briefly explains how scintillators can be directly integrated with devices such as CCD, CMOS or a-Si imagers, which have been already developed for conventional cameras to design x-ray detectors.

A CCD camera uses a small, rectangular piece of silicon to receive incoming light. This silicon wafer is a solid-state electronic component, which has been micro-manufactured and segmented into an array of individual light-sensitive cells called *photosites*. Each photosite is one element of the whole picture that is formed, thus it is called a picture element, or *pixel*. The

CCD photosites accomplish their task of sensing incoming light through the photoelectric effect, which is a characterization of the action of certain materials to release an electron when hit with a light photon. The electrons emitted within the CCD are fenced within nonconductive boundaries, so that they remain within the area of the photon strike. As long as light is allowed to impinge on a photosite, electrons will accumulate in that pixel. When the source of light is extinguished (e.g., the shutter is closed), simple electronic circuitry and a microprocessor or computer are used to unload the CCD array, count the electrons in each pixel, and process the resulting data into an image on a video monitor or other output media. CCD imagers are the established leading technology for high performance optical imaging. Unlike matrix-addressed silicon panels and CMOS imagers, CCDs read out their signal by transporting charge packets across the silicon substrate [4].

Well-designed scintillator systems produce ten or more electrons per detected x-ray and when detected with CCDs can detect 100,000 x-rays per pixel before saturation. If the scintillator is directly coupled to the detector, radiation can penetrate to the semiconductor detector with a small number of events producing large charge and noise.

A fiber optic coupler, as shown in Fig. 4, is used to avoid domination of the signal due to absorption and thereby can be integrated easily with CCDs. The advantage of using CCDs includes high resolution, low noise and increased sensitivity. Scintillators can also be coupled with optics (mirrors and lenses) as shown in Fig. 5. Light produced from the scintillator due to x-ray absorption is reflected from a mirror and focused on to the CCD using a lens. The scintillator to mirror distance and mirror to lens distance plays an important role in deciding the quality of the image produced.

CMOS imagers are matrix-addressed photodiode arrays. They take advantage of a highly developed manufacturing infrastructure by using the same fabrication processes and equipment

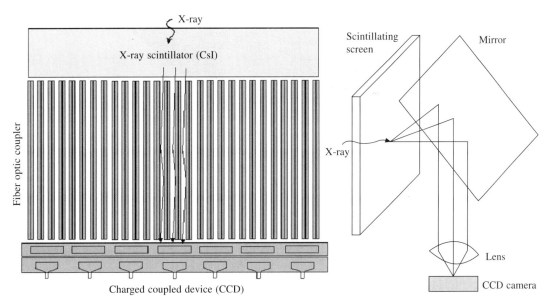

Fig. 4 Integration of CCD to scintillator using fiber optic coupler.

Fig. 5 Integration of CCD to scintillator using lens.

that is used to make microprocessors and logic arrays. In other words, the integrated circuit industry has already paid for the technology development and fabrication equipment, and is continuing to do so. Today's small-line width processes allow the addition of special features on a per-pixel basis, vastly improving the performance of the CMOS array. Noise levels comparable to CCDs can now be achieved, and the dynamic range of a CMOS imager is typically several times larger.

CMOS technology has advantages for integrated cameras, such as low operating power, lower clock voltage and easy integration of analog and digital functions on the chip, in contrast to the more specialized CCD processing technology. Over the last decade, CMOS based image sensors have benefited from the continuing miniaturization and low-power design developments of CMOS technologies. Currently, these trends have improved the quality of CMOS image sensors to a degree acceptable for many imaging applications. In addition, CMOS arrays can integrate timing and readout functions on the same device. Their highly integrated architecture allows the design of a "system on a chip" (SOC), which is ultimately less costly than an imager requiring a large amount of support electronics.

Fig. 6 shows a sensor structure of CMOS containing 2 × 2 pixels. The scintillator guides consist of CsI crystals placed within aluminum cavities [5]. Scintillating crystal converts x-ray photons to light photons which are read by the CMOS photodiode. The aluminum walls surrounding each pixel are highly reflective and guides the visible light to the photodiodes. This eliminates the lateral spread of light to adjacent pixels and thereby reduces blurring.

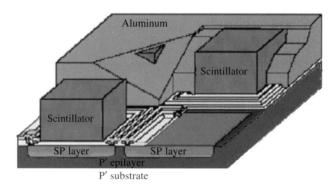

Fig. 6 CMOS sensor structure for 2 × 2 pixels.

3.4 Electronic Readout Process

The readout process can be classified into two categories, viz. readout as in a CCD camera and readout as in a CMOS camera (or) a-Si panels.

3.4.1 CCD

CCDs capture light on the small photosites on their surface and get their name from the way that charge is read after an exposure. A schematic of the CCD assembly in a typical x-ray detector is shown in Fig. 7. To begin, the charges on the first row are transferred to a read-out register. From there, the signals are then fed to an amplifier and then on to an analog-to-digital converter. Once a row has been read, its charges on the read-out register row are deleted. The next row then enters the read-out register, and all of the rows above march down one row. The charges on each

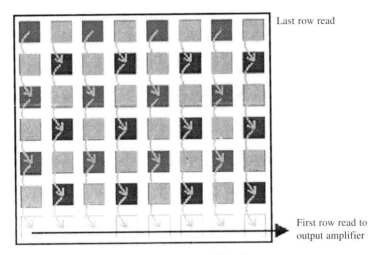

Fig. 7 Readout from CCD pixel array.

row are "coupled" to those on the row above so when one moves down, the next follows to fill its old space. In this way, each row can be read one row at a time.

3.4.2 CMOS

Pixels using photodiodes as optical elements are the most popular of the various possible techniques for constructing a pixel array. Such photodiode-based CMOS image sensors come in two architectures: passive pixel sensors (PPS) and active pixel sensors (APS). The CMOS PPS sensors architecture uses single charge amplification functions for an entire array of pixels. On the other hand, CMOS APS devices incorporate charge amplification units into each individual pixel, as shown in Fig. 8. CMOS APS have a lower fill factor (defined as the pixel area that is photo-sensitive) compared to PPS due to space penalties for extra in-pixel amplification transistors. However, this is outweighed by better noise performance. This makes CMOS APS the most commonly used pixel architecture and is used in x-ray detectors.

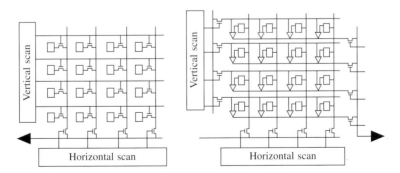

Fig. 8 Readout from CMOS pixel array—PPS and APS.

In a CMOS pixel array, each photosite contains a photodiode that converts light to electrons, a charge-to-voltage conversion section, a reset and select transistor and an amplifier section.

Overlaying the entire sensor is a grid of metal interconnects to apply timing and readout signals, and an array of column output signal interconnects. The column lines connect to a set of decode and readout (multiplexing) electronics that are arranged by column outside of the pixel array. This architecture allows the signals from the entire array, from subsections, or even from a single pixel to be readout by a simple X-Y addressing technique.

Amorphous silicon imagers are typically photodiode arrays with an active thin-film transistor (TFT) matrix readout, similar to active matrix flat panel displays. The thin-film technology enables very large imaging areas (20 cm × 20 cm and larger), and the amorphous silicon has very high radiation tolerance. However, it has limitations in both resolution (typically >100 μm) and performance (noise, contrast) owing to the less-than ideal properties of the amorphous silicon semiconductor. In addition, it requires a specialized fabrication process with a dedicated manufacturing facility, increasing both production and development costs relative to competing technologies.

4. Digital Detectors for X-ray Imaging

Digital detectors can be broadly classified into three major categories, namely, image intensifiers, computer radiography and digital radiography using flat panels.

4.1 X-ray Image Intensifiers

X-ray image intensifier (XII) based radiography is referred, as real-time radiography or fluoroscopy or radioscopy, which allows real-time visualization of sequence of radiographic images with a speed of 30-100 frames/sec. The x-ray image intensifier converts the transmitted x-ray photons through object into a brightened, visible light image on a TV monitor. Because of its speed it is widely used for an on-line inspection of castings in automotive industries and pipes in pipe mills. It finds application in semiconductor industries for high magnification radiography of miniature components using micro-focus x-ray sources.

4.1.1 Construction and Operation

The construction of x-ray image intensifier (XII) is shown schematically in Fig. 9 [7]. The XII consists of an input phosphor which converts the x-ray photons to light photons. Cesium iodide

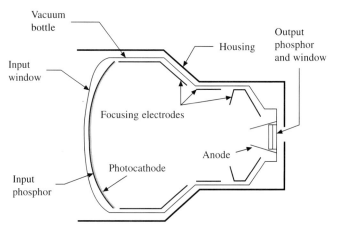

Fig. 9 Construction of x-ray image intensifier (XII).

activated with sodium (CsI : Na) is the most widely used scintillator material for input phosphor screens as it has good absorption efficiency for x-rays. Gadolinium oxy-sulphide is the other phosphor used. Currently, the thickness of an input phosphor layer typically measures between 300 and 450 mm, depending on the image intensifier type and technology used. The thickness of the input phosphor layer is a compromise between spatial resolution and x-ray absorption efficiency. The lateral diffusion of light photons limits the spatial resolution of detector with thicker scintillator layer. This limitation is overcome by using CsI : Na with needle structure. The light photons are then converted into photoelectrons by photocathode layer next to phosphor layer. The photocathode layer is made of antimony-cesium (SbCs3). To maximize the conversion efficiency from light photon to photoelectron, light emitted from the input phosphor should match the sensitivity spectrum of the photocathode.

The accelerated photoelectrons are focused down to the size of the output phosphor by a series of electrostatic focusing electrodes. The output phosphor screen converts the accelerated electrons into light photons that may be captured by various imaging devices. Changing the voltage applied to the electronic lenses inside an image intensifier will change the magnification mode of the image intensifier. In a magnification mode, a smaller area of the input phosphor is used, giving the effect of zooming in on the image. Because the input field size is reduced, the exposure to the input phosphor must be increased to maintain a constant brightness level at the output phosphor. In fact, to maintain the same noise level, the dose quadruples when the magnification is doubled. Charged-coupled devices are used as read-out camera in modern image intensifiers. The input phosphor, photocathode, electrodes and output phosphor screens are enclosed in a vacuum bottle. XIIs are available with various field of view of 4' or 22' in diameter. The imaging performance of XIIs is limited by a variety of imperfections or artifacts including lag, vignetting, veiling glare, pincushion distortion, and S-distortion [7].

4.1.2 Physical Characteristics
Brightness gain, conversion factor, contrast ratio and spatial resolution characterize an image intensifier [7].

(a) Brightness gain
The gain in image brightness in XII results from the combined effects of image minification and the acceleration of the electrons. The minification gain is defined as the ratio of input area to the output area of the image intensifier. Because of minification the number of photoelectrons per unit area at the output phosphor increases. The minification gain does not improve the statistical quality of the fluoroscopic image, as the total electrons remain the same. It will not change the contrast of the image, but it will make the image appear brighter. Flux gain is defined as the number of photons generated at the output phosphor for every photon generated at the input phosphor. The flux gain results from the acceleration of photoelectrons to a higher energy so that they generate more fluorescent photons at the output phosphor. The total brightness gain of the image intensifier is the product of minification gain and flux.

(b) Conversion factor
The conversion factor is defined as the output luminance level of an image intensifier divided by its entrance exposure rate. It is a measure of how efficiently an image intensifier converts the x-rays to light. Conversion factors have units of candela per square meter per milliroentgen per second ($[cd/m2]/[mR/sec]$). The higher the conversion factor, the more efficient is the image intensifier.

(c) Contrast ratio
Contrast ratio is a measure of veiling glare and it is measured in two ways, viz. large area and small detail contrast factor. The contrast ratio of an image intensifier is defined as the brightness ratio of the periphery to the center of the output window when the center portion of an image intensifier entrance is totally blocked by a lead disk. The large area or 10% area contrast ratio is measured by putting a lead disk, which has a surface area equal to 10% of the useful entrance area of the image intensifier, at the center of the input surface of the image intensifier. The small detail, or 10 mm area contrast, is measured by putting a 10 mm lead disk at the center of the input surface of the image intensifier.

The other performance measurements of XIIs are limiting spatial resolution, modulation transfer function, brightness uniformity and distortion. The conversion factor and spatial resolution can be measured together using a metric called detective quantum efficiency (DQE) (see Section 5 for details).

4.1.3 Artifacts in x-ray Image Intensifier Tubes

(a) Veiling glare
Contrast of image in XIIs is affected by scattering of x-rays, light and electrons with in the image intensifier tube. This is called veiling glare. Veiling glare degrades object contrast at the output phosphor of the image intensifier. Contrast ratio is a good measure of determining the veiling glare.

(b) Vignetting
XII images of a uniform object are generally brighter in the center than in the periphery due to an unequal brightness gain in different regions of the field of view. This effect is also called vignetting.

(c) Lag
Lag or hysteresis is due to persistence luminescence of light in the scintillator after x-rays are stopped. This effect is very less in modern day image intensifiers.

(d) Pincushion distortion
It occurs due to geometric, nonlinear magnification across the image. The magnification difference at the periphery of the image results from the projection of the x-ray beam onto a curved surface. The distortion can be easily corrected by using post processing techniques.

(e) S-distortion
Due to external electromagnetic sources the electron paths at the perimeter of the image intensifier can be affected more than those nearer the center. This characteristic causes the image XII system to distort with an S shape. Larger image intensifiers are more sensitive to the electromagnetic fields that cause this distortion. In modern XIIs the electron path is shielded from external magnetic field using Mu-metals.

4.1.4 Advantages and Limitations
Faster read-out time, conversion efficiency and low cost are the major advantages of XIIs. The imaging performance of XIIs is limited by variety of imperfections or artifacts including lag,

vignetting, veiling glare, pincushion distortion, and S-distortion. XIIs are also very bulky and fragile and hence cannot be used for field applications.

4.1.5 Applications

X-ray image intensifiers are used in electronic industries for high definition radiography of integrated circuits and printed circuit boards using micro-focal x-ray sources. XIIs are also used in x-ray testing machines, which are dedicated systems for inspection of castings in automotive industries and welds in pipe mills. High-speed image acquisition capability of XIIs can be used for acquiring data for volumetric computed tomography. However, spatial resolution of this system is limited. XIIs are used in security imaging for scanning baggages and cargos in airports. They are also used in food industry for screening contaminants in foods.

4.2 Computed Radiography

Computed radiography (CR) systems using storage phosphors have continued to evolve in parallel with instant readout digital radiography systems. CR refers to x-ray imaging using imaging plates containing storage phosphors as detection medium. The latent image formed in the imaging plate after radiation exposure is scanned using a laser beam to get digital radiographic image of object exposed. CR is getting popular in non-destructive evaluation of materials due to its large dynamic range, portability and faster digital imaging capability as compared to conventional film based radiography.

4.2.1 Storage Phosphors

Photo stimulated luminescence (PSL) is the phenomenon responsible that makes imaging using storage phosphors possible [6]. The phosphors used are most often from barium flurohalide family deposited on to substrate to form imaging plate. Unlike the phosphors used in instant readout digital radiography system, in storage phosphors the optical signal is not derived from prompt response to incident x-ray radiation. The optical signal is obtained by releasing trapped charges in metastable states using external optical stimulation. This process of stimulating luminescence in storage phosphors using external optical photons is called photo stimulated luminescence.

The photostimulable phosphor first used for CR was $BaFBr : Eu^{2+}$. $BaFBr : Eu^{2+}$ is a good storage phosphor as it can store latent image for a long time, e.g. the latent image 8 h after irradiation will still be ~75% of its original information. The family of phosphors $BaFX : Eu^{2+}$, where X can be any of the halogens Cl, Br or I. The decay time after photostimulation of all these phosphors is now known to be approximately the same (~0.7 μs) and so they can all be used in CR. Physical properties of phosphors [6] used in CR and DR is given in Table 2.

The spectrum of light emitted by an efficient phosphor is controlled by a dilute (<1 atomic %) impurity called an *activator*. Such activated phosphors have a characteristic line spectrum caused by the isolated atom in the host or *matrix*. In BaFX phosphors used in CR the activator is Eu^{2+}, which substitutes for Ba in the crystal lattice. Additionally, in a photostimulable phosphor there should be effective electron and hole traps at every activator site so that the maximum number of x-ray induced excitations can be trapped. In $BaFBr: Eu^{2+}$ the electrons are trapped at positive ion (Br or F) vacancies, forming an F-center. The energy difference between the electron traps and the conduction band edge must be small enough to allow stimulation with laser light, yet sufficiently large to prevent significant random thermal release of the charge carriers from the traps.

Table 2. Physical properties of conventional and photostimulable phosphors

Phosphor	Z	Ek (keV)	Eg (eV)	Density (g/cc)	W (eV)	G (Photons/ 50 keV)	Decay time (μs)	Light emission peak (nm)	Spectrum for stimulation (nm)
$Gd_2O_2S:Tb^{3+}$	64	50.2		7.34	20	2500	~3	550	na
$BaFCl:Eu^{2+}$	56	37.4	(~8)	4.56	25	2000	0.7	390	500-600
$BaFBr:Eu^{2+}$	56	37.4	8.3	5.1	360	140	0.7	390	500-650
$BaFBr_{0.85}I_{0.15}:Eu^{2+}$	56/53	37.4/33.2	(~8)	(5.1)	360	140	0.7	390	550-700
$BaFI:Eu^{2+}$	56/53	37.4/33.2	(~8)	(~5.6)			0.6	405	550-700
$CsI:Tl^+$	55/53	36/33.2	6.2	4.52	20	2500	0.98	550	na
$CsBr:Eu^{2+}$	55	36	7.3	4.45	250	200	0.7	440	685
$RbBr:Tl^+$	37/35	15.2/13.4		3.35			0.35	433	735

During exposure to radiation, Eu^{2+} molecules in imaging plates release electrons. The free electrons are trapped in F-center, which are halogen ion vacancies in crystal as shown in Fig. 10. The trapped electrons are released after subsequent exposure to visible laser radiation. The released electrons return to the europium (Eu^{3+}) molecules and emit the luminescence at 390 nm [6]. The Eu^{3+} molecules are converted back to Eu^{2+}.

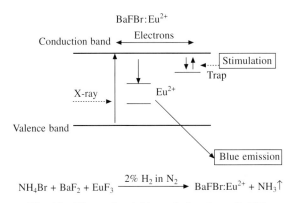

Fig. 10 Photostimulable emission from BaFBr.

4.2.2 Imaging Plates and Cassettes Construction

Imaging plates in general consist of a protective layer, a phosphor layer and a support plate usually made of polythene. The construction of imaging plates varies from manufacturer to manufacturer. The protective layer protects phosphor layer from mechanical wear and it is immune to chemical cleaning solutions. The phosphor layer consists of storage phosphors, coated on support layer using adhesives. There will be an anti-halo or light reflection layer in between phosphor and support layers in imaging plates. Its function is to protect reflection of laser light and allows only stimulated light. The typical construction of a imaging plate [8] is given in Fig. 11.

The imaging plates need to be placed in cassettes before exposure. The typical construction of cassette for NDE application is shown in Fig. 12. It consists of multiple layers, cassette cover, front lead screen, magnetic holders and lead foil at the back. The lead screen helps to reduce the scattered radiations. The magnetic holders maintain closer contact between the phosphor and lead screens.

4.2.3 Scanning Systems

CR systems used for NDE applications are cassette-based systems [6] where the imaging plate is enclosed in a light-tight cassette for the x-ray exposure, and subsequently moved by hand to the readout system. A typical read-out system is shown in Fig. 13. It is usually a flying

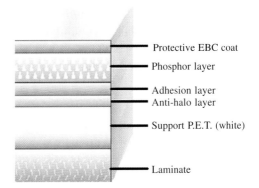

Fig. 11 Construction of a typical imaging plate.

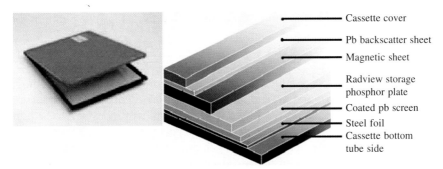

Fig. 12 Typical construction of cassette for NDE application.

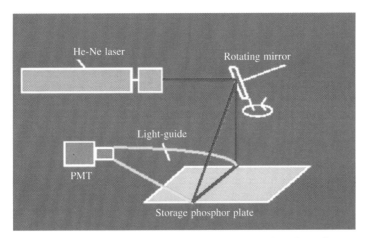

Fig. 13 Typical read-out system from storage phosphor plate.

spot read-out system, i.e. a laser spot is scanned with a mirror over the exposed imaging plate in a point-by-point raster pattern. For mirror scanning, highly collimated and circular beams with Gaussian profile is desirable. This can be accomplished using helium-neon gas lasers or solid-state laser diodes. The blue light emitted by imaging plate is sampled by using a photomultiplier tube and the signal is amplified and digitized to form the image.

4.2.4 Advantages and Limitations

Imaging plates are available in multiple sizes in the market to cover wide range of applications. Large-area plates are conveniently produced, and because of this format, images can be acquired quickly. The plates are reusable, have linear response over a wide range of x-ray intensities, and are erased simply by exposure to a uniform stimulating light source to release any residual traps. It is flexible and can be wrapped around any object shape and gives extremely sharp images. Imaging plates are portable which enables field radiography. It is cost effective and needs significantly lesser exposure time as compared to film radiography.

As compared to 2D array digital radiography panels, imaging plates require more exposure and scanning time to acquire images. CR offers productivity benefit better than film, however, it is not comparable to digital radiography. Other limitation of CR is that because the traps are located throughout the depth of the phosphor material, the laser beam providing the stimulating light must penetrate into the phosphor. Scattering of the light within the phosphor causes release of traps over a greater area of the image than the size of the incident laser beam. This results in loss of spatial resolution. This limitation can be overcome by using phosphors in the form of needles as used in digital radiography panels.

4.2.5 Applications

Computed radiography has the potential to replace conventional film based radiography for many applications in shop floor and in field radiography. Imaging plates require lesser exposure time as compared to film and also do not involve any chemical processing to acquire images. Formation of digital images offers other advantages of post processing and archival of images. Imaging plates are continuously being under development to match image quality of different types of high contrast and high-speed industrial x-ray films. The portability of imaging plates allows carrying out radiography in field and inaccessible areas. This enables corrosion detection and integrity of process pipes in petrochemical industries and power plants. It is also being explored in aerospace industries for site radiography to inspect airframes and other composite structures. It can be used to inspect concrete structures using high-energy radioisotopes. It has a wide dynamic range as compared to film, which has an advantage of inspecting castings in automotive industries. This can be also used for security imaging for inspection of suspicious packages in airport and crowded areas.

4.3 Digital Radiography Using Flat Panel Detectors

As discussed in the previous sections, there are many digital x-ray imaging systems available today and continuous improvements are being made on these technologies. Among these systems, the flat panel detectors are the most attractive digital imaging system currently available. Flat panel based system are usually faster and have a very high dose efficiency. In a typical industrial application a few milliseconds of exposure is enough to record a good image. Different comparative studies have been reported in the literature for the flat panel detectors. These detectors are found to be 10-100 times more dose efficient for many applications.

On-line inspection of machined components or manufacturing parts with reliable automated inspection system at a high throughput is an ideal inspection system. With the availability of flat panel detectors this requirement is achieved. Elimination of chemical processing or a post-processing step like scanning the imaging plate with a laser is avoided in the flat panel detector system. The pixel resolutions are also continuously improving. Commercially, today we can buy off-the-shelf flat panel detectors with 100 microns pixel pitch. Direct recording of the image in the digital form offers several advantages ranging from easy archival to automated inspection and several image processing techniques to enhance the image quality. Automated inspection and ability to digitally enhance the images have reduced operators' fatigue considerably and the x-ray inspections are more reliable with these kind of detectors. Automation has also removed subjectivity or operator's reproducibility issues from x-ray image analysis. Flat panel detectors have brought a revolution in the area of computed tomography imaging. Volumetric computed tomography (VCT) has improved 3D inspection time and quality of the volumetric images.

While flat panel detectors offer many advantages the cost of these detectors is still high compared to other forms of x-ray image recording devices. The initial investment to setup a flat panel detector for x-ray imaging is high but over a period of time the benefits associated with it makes that investment worth. Operational costs associated with these detectors are also projected to be low compared to other forms of x-ray imaging devices.

4.3.1 X-ray Conversion in Flat Panel Detectors

X-ray conversion techniques in a flat panel detector can be broadly classified into two categories, viz. indirect and direct conversion techniques. Indirect conversion technique has two stages of conversion. The first stage is to convert x-ray photons to light photons in the visible range, which is performed using a scintillator, and the emitted light photons can be converted to a digital image, using one of the techniques discussed in Section 3.3. X-ray conversion to digital electrical signal is shown schematically in Fig. 14.

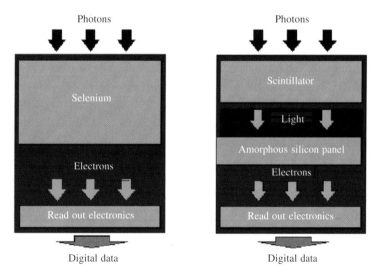

Fig. 14 X-ray conversion techniques—direct and indirect conversion.

When radiation is incident on scintillators, electrons get excited to higher energy states. The electron in the higher energy state will return back to the lower energy state and this process is referred to as deactivation. Deactivation is usually accompanied with the release of light photon (in the visible range).

Direct conversion detectors consist of an x-ray photoconductor layer (typically amorphous selenium), transforming x-ray photons into electric charges, grown directly on top of the TFT (thin film transistor) charge collector and readout layer. Amorphous selenium is used because of its excellent x-ray detection properties. Direct conversion of x-rays to electrical signal also offers very high spatial resolution.

An electric field is applied across the selenium layer before exposition to x-rays. X-ray exposition generates electrons and holes within the selenium layer as shown in Fig. 15. Due to the electric field, these charges migrate nearly perpendicularly to both surfaces of the selenium layer, without much lateral diffusion. At the bottom of the layer, charges are drawn to the charge-collection electrodes, where they are stored until readout. During the readout, the charge of the capacitors of every row is conducted by the transistors to the amplifiers [2].

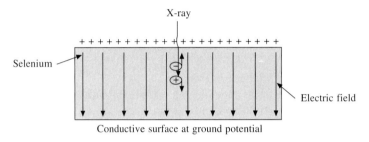

Fig. 15 Direct conversion using selenium scintillator.

Although direct conversion is a more favorable way of forming x-ray images and amorphous selenium does provide a good x-ray conversion efficiency in that class of materials, the dose efficiency of these detectors are still not comparable to indirect conversion detectors, for example, detectors using cesium iodide scintillator and amorphous silicon photodiodes. The quantum efficiency of the indirect conversion technique has an advantage over the direct conversion technique although it may not compete with the resolution capability. The extra stage of x-ray photon conversion to a photon in the visible range, which is the main reason for the resolution loss, is addressed by using needle-like cesium iodide crystals. Needle structure of the crystal helps in keeping the collimation of the light photons in the cesium iodide layer. Light scattering effect, sensitivity and memory effect following strong exposure (hysterisis) are some of the limitations of indirect conversion technique.

4.3.2 *Electronics of Flat Panel Detectors*

Doped amorphous silicon (a-Si:H) transistor-addressed arrays have demonstrated that they are the leading technology for flat panel imaging. Processing of a-Si : H allows for deposition on large substrates, making them ideal for projection imaging. Imagers on a single substrate with up to 2304 × 3200 pixels (29.2 × 40.6 cm) at 27 μm pitch have been demonstrated and are in production.

In an a-Si : H imager, each pixel has a TFT as shown in Fig. 16. The sensing element for light is a reverse biased PIN diode. The PIN diode has heavily doped p-type and n-type regions separated by an intrinsic region. When reverse biased, it almost acts like an constant capacitance and when forward biased it behaves as a variable resistor. This has sufficient capacitance to store charge during the frame time. As the charge accumulates, the voltage at node A approaches the bias voltage, and is reset after readout to restore the reverse bias on the sensor. All gates of TFTs within one row are connected to a gate driver circuit as shown in Fig. 17. When one row of gates is driven high, charge stored within all pixels of that row is released. The source contacts for all

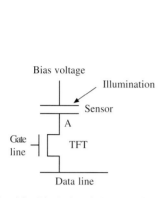

Fig. 16 Pixel circuit in an a-Si:H detector.

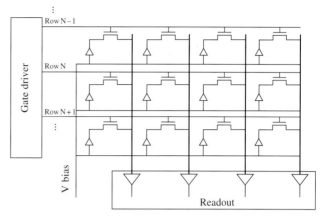

Fig. 17 Flat-panel-pixel architecture.

TFTs within one column share one data contact, which is connected to a charge sensitive amplifier. X-ray sensitivity is obtained by coupling to a phosphor screen. Typically a Gdz02S : Tb phosphor is placed in contact with the array, although CsI has also been deposited on a-Si : H imagers to achieve high-resolution imaging.

4.3.3 Performance

X-ray imaging detectors are usually expected to have certain properties, which are related to image quality and performance of the detectors for various applications. These include linearity, dose sensitivity, spatial resolution, dynamic range and minimum artifacts.

(a) Linearity

Linearity of detector response is an important factor in producing high-quality digital radiographic images. All digital detectors exhibit some degree of spatial inhomogeneity in their x-ray response. In particular, slight variation in pixel-to-pixel sensitivity is usually observed. To produce a smooth image these variations must be compensated. This correction is commonly referred to as flat fielding or normalization, and is typically accomplished by applying a linear transformation on a pixel-by-pixel basis to the raw image data, using offset and gain calibration data. For operating the normalization procedure over a wide range of exposure conditions, the detector's basic response should be linear over the useful dynamic range. Fig. 18 shows a plot of the linearity of a typical flat panel detector system. A relative exposure is plotted on the x-axis and the gray value counts obtained from the detector are shown along the y-axis. The detector is strictly linear with the incident dose.

(b) Dose sensitivity

Lower dose sensitivity, i.e. obtaining a significantly large signal level at a lower dose is an ideal condition for both industrial and medical applications. While in medical application lowering the dose avoids radiation hazards to the patients, in industrial scenario lowering the dose improves productivity and also minimizes radiation damage to the semiconductor and phosphors used in the detectors and hence improves its life. Detector quantum efficiency (for details, see Section 5) is a metric to quantify image quality in terms of dose irradiated to the detector. Detectors should

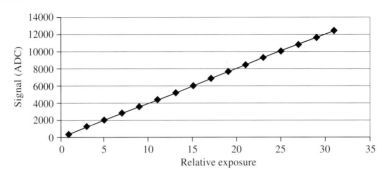

Fig. 18   Linearity of digital detectors.

have as high as possible quantum efficiency for better image quality at lower dose. In general, the modern days flat panel detectors shows higher quantum efficiency than other forms of available x-ray imaging detectors.

All images generated by quanta are statistical in nature, i.e. although the image pattern can be predicted by the attenuation properties of the material, it will fluctuate randomly about the mean value. The radiation sources are characterized by its Poisson nature of photon distribution in time. Hence, the noise level is seen to be proportional to the square root of the mean signal level. The overall noise that we see in a digital x-ray image is the cumulative effect of many noise sources such as radiation source, noise generated during interaction with the materials and electronic noise. Usually the overall noise behavior also follows Poisson statistics and indicates that the noise is quantum limited. Noise is also directly related to dose efficiency of the detector. While the detector industries are striving to produce high quantum efficiency detectors to improve noise performance, there are other options as well to minimize the noise since the images are discretized and available in digital format.

There are two ways by which the Poisson distribution of noise can be taken as an advantage to improve the quality of the image. When the x-ray tube current is increased, the number of photons generated will increase, and hence, the mean count on the detector will increase. As long as the saturation of the detector is not reached, this will enable reducing relative noise in the image.

The second method is to acquire a series of images in time domain and average them. The information at each pixel with time is expected to be a Poisson distribution. Obtaining this information over a period of time and estimating the mean of the distribution at each pixel will provide a noiseless image without introducing artificial blurring in the image.

(c) Spatial resolution

Higher spatial resolution is required for better clarity on small size features. The ability to resolve the minimum feature size in an image is usually measured in terms of spatial resolution. Digital detectors have finite size pixels where the x-ray energy quanta are discretized to electrical signal. Each of these pixels is a basic building block for the image. The pixel size determines the resolution capability [4] for the image-recording device. Spatial resolution of the system is not only limited by the detectors; in fact the imaging system as a whole decides the spatial resolution of the system. Focal spot of the x-ray source and magnification plays a role in further degrading the spatial resolution offered by the detecting device. While magnification is an attractive imaging technique to resolve small features, finite focal spot size of the x-ray source creates its own problem in terms of geometrical unsharpness and blurring the image. Combination of micro focal x-ray source and high dose efficiency digital detectors is a great imaging device since we can reduce geometrical unsharpness considerably through the miniaturization of the x-ray focal spot. High dose efficiency of the digital detector is an advantage for micro focal x-ray tubes as these are usually low flux tubes and hence otherwise would require a large exposure time to get a reasonable image quality. Fig. 19 shows some magnification radiography imaging taken by combination of a micro focal x-ray tube and a flat panel detector system. Image quality at 50 × magnification is excellent due to negligible geometrical unsharpness.

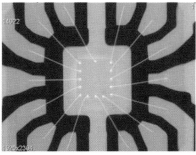

Object : IC
Kv : 70
mAs : .5
FS : 10 microns
Mag. : 50×

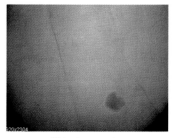

Object : ceramics
Kvp : 70
mAs : .5
FS : 10 microns
Mag. : 50×

Object : IC
Kv : 70
mAs : .5
FS : 10 microns
Mag. : 50×

Fig. 19 Magnification images using a micro focal x-ray tube. The object shown is a IC chip on the left side and on the right side a ceramics material is magnified 50 times to see a defect.

Although, as traditionally viewed, digital detectors have lower spatial resolution capability than film x-ray imaging and is actually not a valid assumption. It is possible to design a digital radiography system with equal or more spatial resolution capability by combining micro focal x-ray tubes with digital detectors and operating at a large magnification. The down side of this arrangement is that we compromise in the field-of-view. Field-of-view decreases with increase in magnification.

(d) Dynamic range
A wider dynamic range allows more usable area of the detector response range. For example, in the x-ray film case and the dynamic range is limited by the area of the response curve where the optical density is typically between 1.5 and 2.5. Below an optical density of 1.5 the signal is unreliable and above 2.5 it is too intense. In contrast, for a digital detector using a 12-bit analog to digital converter, we can have gray values ranging between 0 and 16383. For gray value of 1000 onwards the visibility is clear. Hence, we get a very large usable portion of the response

curve for image formation. Wider dynamic range of the detector system helps cover more variation of depth of the object material (latitude coverage) using a single exposure.

(e) Artifacts

Different forms of image distortions are possible using different detectors. Specifically digital detectors have the following types of image distortions.

Distortion: Image intensifier has distortion as seen in Fig. 20 because of the projection of the x-ray image onto the curved input phosphor. Since no projection is involved in flat panel detectors and the sensitivity of the detector is uniform throughout, these distortions are absent.

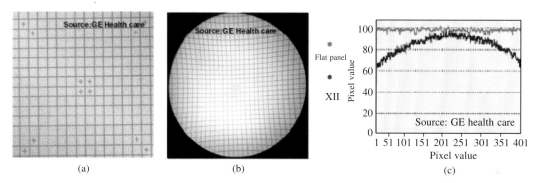

Fig. 20   (a) No distortion seen in flat panels, (b) distortion seen in image intensifiers and (c) comparison of brightness uniformity between flat panels and image intensifiers.

Blooming: A CCD camera records images by converting photons of light into electrons. The electrons are temporarily stored in individual picture elements (pixels) on a photosensitive detector or chip. At the end of an exposure, the accumulated charges are read off the chip and sent to the computer for conversion to an image. During the long exposures needed for thick objects, thin objects in the field can exceed the full well capacity (the electron-holding capacity) of the pixels on which they are being recorded. As a light-gathering pixel exceeds its capacity to hold captured photons, the excess energy spills over into the adjacent pixel (or pixels, if the second pixel also fills to its capacity). These blooming artifacts are absent in flat panels.

5. Image Quality

Quantifying image quality from an image system is essential for performance comparison. With several different types of image recording devices available in the market, users need to evaluate the correct imaging device for the application they are targeting. Image quality can be defined completely in the image domain itself, i.e. without relating it to the imaging system. Such image quality parameters are important for image processing applications. Image quality during data acquisition is more important from a system perspective. It is important to use a single metric to quantify image quality from both the system and the image domain. Specific to an industrial digital radiography imaging system, an ideal imaging system, is one, which can provide maximum resolution at a lower dose to the detector and highest contrast-to-noise ratio in the acquired image data.

Measurement of resolution capability, noise performance and ability to provide a better image at a lower dose to the recording device needs to be quantified. These quantities are mainly related to the image quality performance of the imaging system. Modulation transfer function (MTF) is usually a popular measure for the resolution capability. Limiting spatial resolution (LSR) is another metric that is often used to quantify resolution of an imaging system. Noise is characterized by noise power spectrum (NPS). Noise in digital x-ray imaging system is mainly related to radiation dose. More the radiation dose less is the observed noise. Noise is quantum limited and is predominantly due to photon arrival statistics [10]. Better image quality requires higher dose, but is not favorable for detector life. Radiation damage is severe in digital detectors and hence dose irradiated to the detector needs to be minimized. Hence, this becomes an optimization problem, where we target to get an acceptable image quality at a minimum dose to the detector. This factor is often characterized by a quantity known as detector quantum efficiency (DQE). Better DQE system leads to a better image quality in the image data, which is usually measured as contrast-to-noise ratio (CNR) and also improves the detector life.

5.1 Resolution of an Imaging System

The details of feature in an image are an integral part of its appeal. We spend a great deal of time, energy and money acquiring equipment to get sharp images. The sharpness of an imaging system or of a component of the system (x-ray source, recording device, geometrical distance) is characterized by a parameter called the modulation transfer function (MTF), also known as spatial frequency response. We present here a visual and mathematical explanation of MTF and how it relates to image quality. A sample is shown in Fig. 21. The top is a target composed of bands of increasing spatial frequency on the plane of the recording device. Below the cumulative effect of the imaging system on the pattern is shown. The imaging system is poor in resolution capability as we move downward in Fig. 21. This kind of bar pattern is usually employed to access the resolution performance of the imaging system. The bar pattern is usually represented by the spatial frequency and expressed as dots per inch, pixels per line, lines per millimeter etc. The bar patterns form the basis of defining resolution of x-ray imaging system.

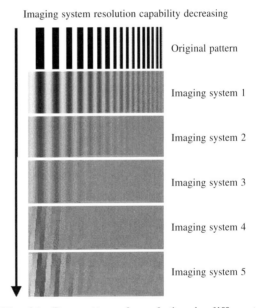

Fig. 21 Bar pattern degradation in different imaging systems.

5.2 Limiting Spatial Resolution

Limiting spatial resolution (LSR) is a measure of the spatial frequency at which one can no longer see a high-contrast, structured periodic test pattern under the most favorable test conditions—for instance, in the absence of scatter and focal spot penumbra and with the use of high radiation dose. Usually LSR is measured with high-contrast objects (lead) and at a very high dose at a magnification close to one (to avoid geometrical unsharpness). The measurement procedure is straightforward. Once the image data is acquired, a trained eye will examine the image and take a decision up to which spatial frequency the pattern is discernable. LSR is measured as line pairs per mm.

5.3 Modulation Transfer Function

Modulation transfer function (MTF) is the spatial frequency response of an imaging system or a component; it is the contrast at a given spatial frequency relative to low frequencies [11-15]. Spatial frequency is typically measured in cycles or line pairs per millimeter (lp/mm), which is analogous to cycles per sec (Hertz) in audio systems. High spatial frequencies correspond to fine image detail. The more extended the response, the finer the detail and the sharper the image. Most of us are familiar with the frequency of sound, which is perceived as pitch and measured in cycles per sec (Hertz). Audio components like amplifiers and loudspeakers etc. are characterized by frequency response curves. MTF is also a frequency response, except that it involves spatial frequency cycles (line pairs) per distance (millimeters or inches) instead of time. The response of imaging system components (focal spot size, magnification, pixel pitch) tends to roll off at high spatial frequencies. These components can be thought of as low pass filters that pass low frequencies and attenuate some of the high frequencies.

The essential meaning of MTF is rather simple. At frequencies where the MTF of an imaging system or a component is 100%, the pattern is unattenuated and it retains full contrast. At the frequency where MTF is 50%, the contrast at half its original value is retained and so on. MTF is usually normalized to 100% at very low frequencies. Contrast levels from 100% to 2% are illustrated in Fig. 22 for a variable frequency sine pattern. Contrast is moderately attenuated for MTF = 50% and severely attenuated for MTF = 10%. The 2% pattern is visible only because viewing conditions are favorable since it is noiseless. It could easily become invisible under less favorable conditions.

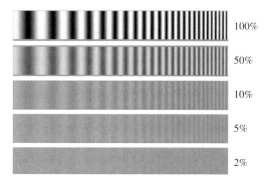

Fig. 22 Bar pattern visualized at different modulation levels.

A typical MTF curve is shown in Fig. 23. As the spatial frequency increases, the modulation in the contrast to the specific frequency drops. For example, at a spatial frequency of 2.5 lp/mm there is a 20% modulation of the contrast.

MTF response of an imaging system is a cumulative effect contributed by each of its subsystems. In a typical x-ray imaging system the MTF response is expected from two major subsystems, the focal spot of the x-ray source and the x-ray recording device. Assuming a digital x-ray recording device, the MTF response can be further classified due to the finite pixel size and due to the diffusion of light.

5.3.1 Modulation Transfer Function of a Detector

The MTF response of a detector can be derived through the line spread function (LSF) of the detector [11-15]. If an x-ray beam is passed through a very narrow metal slit and the resulting 'line' image is recorded on the detector, the response at the detector will be as shown in Fig. 24 (a). This is the associated spread for a thin line and called as the LSF. If a test object, for example a bar pattern is imaged, an ideal imaging system will produce a profile as shown in Fig. 24 (b).

Digital Radiography for Non-destructive Testing 201

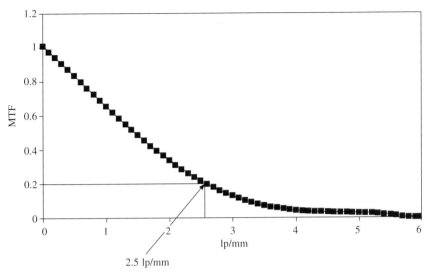

Fig. 23 A typical MTF response curve of a digital x-ray detector.

However, due to the associated LSF characteristics of the system the profile in real life imaging would look like the distorted profile in Fig. 24 (b) in which we can clearly see the blurring of the edges and the density difference between the edges of the bar and the background has decreased.

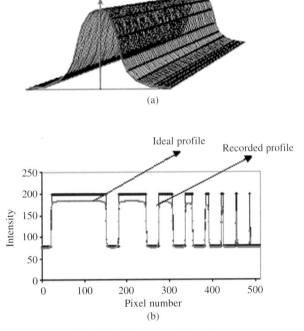

Fig. 24 Line spread function.

Let us consider the case of a one-dimensional transmitted fluence $N(x)$ representing a slit of width L as shown in Fig. 25 (a). The image profile is shown in Fig. 25 (b). Since the transmitted image is symmetric about $x = 0$, it can be represented as a cosine series (in general sines and cosines are required) of the form.

$$N(x) = \frac{a_0}{2} + \sum_{n=1}^{\infty} a_n \cos(k_n x) \qquad (5.1)$$

$$k_n = \frac{n\pi}{L}$$

where

$$a_n = \frac{2}{L} \cdot \int_{-L/2}^{L/2} N(x) \cos k_n x \, dx$$

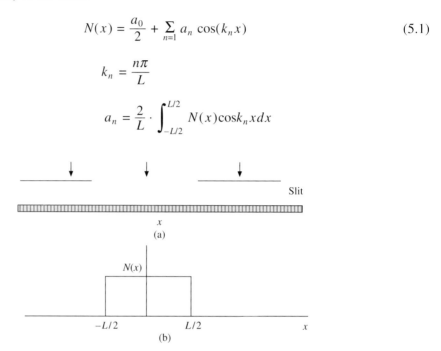

Fig. 25 Fluence through a narrow slit.

Usually, it is more convenient to express the image as an integral over the various spatial frequency components required to represent the image. Any distribution in x can be written as a sum of integrals of the form

$$N(x) = \frac{1}{\sqrt{2\pi}} \int_0^{\infty} \tilde{N}_+(k) e^{ikx} dk + \frac{1}{\sqrt{2\pi}} \int_0^{\infty} \tilde{N}_-(k) e^{-ikx} dk \qquad (5.2)$$

This is a sum over spatial frequencies similar to the discrete example shown in Eq. (5.1), but now with a continuous distribution of frequency components. The quantities $\tilde{N}_+$ and $\tilde{N}_-$ are in general complex numbers and are basically the weighting coefficients for the various frequencies analogous to the discrete points. Eq. (5.2) can be reduced to a form

$$N(x) = \frac{1}{\sqrt{2\pi}} \int_{-\infty}^{\infty} \tilde{N}(k) e^{ikx} dk \qquad (5.3)$$

It can be interpreted that the appearance of negative spatial frequencies in Eq. (5.3) is a reflection of the fact that there must, in general, be terms of the form e^{-ikx}. The artificial concept of negative spatial frequencies just arises in association with writing the integral in compact form. Assuming $N(x) = 1$ when x is in the range of $-L/2$ to $L/2$ and $N(x) = 0$ outside:

$$\tilde{N}(k) = \frac{1}{\sqrt{2\pi}} \int_{-L/2}^{L/2} e^{-ikx} dx = \frac{1}{2\pi} \cdot \frac{1}{ik} [e^{ikL/2} - e^{-ikL/2}] = \frac{L}{\sqrt{2\pi}} \cdot \frac{\frac{1}{2i}[e^{ikL/2} - e^{-ikL/2}]}{kL/2} \quad (5.4)$$

Eq. (5.4) can be further simplified as

$$\tilde{N}(k) = \frac{L}{\sqrt{2\pi}} \cdot \frac{\sin(kL/2)}{kL/2} = \frac{L}{\sqrt{2\pi}} \frac{\sin(\pi f_x L)}{\pi f_x L} = \frac{L}{\sqrt{2\pi}} \cdot \text{sinc}(f_x L) \quad (5.5)$$

Using a narrow slit approximation the line spread function associated at the detector plane due to fluence through a narrow slit can be represented as

$$\text{LSF}(x) = N(x) = \frac{1}{\sqrt{2\pi}} \int_{-\infty}^{\infty} \delta(k) \cdot M(k) \cdot e^{ikx} dk \quad (5.6)$$

where $M(k)$ is the system transfer function associated with the detector. $M(k)$ is defined as

$$M(k) = \int_{-\infty}^{\infty} N(x) \cdot e^{-ikx} dx \quad (5.7)$$

A general formula for a normalized system transfer function may be obtained by normalizing to the zero frequency value and can be written as

$$M(k) = \frac{\int_{-\infty}^{\infty} \text{LSF}(x) \cdot e^{-ikx} dx}{\int_{-\infty}^{\infty} \text{LSF}(x) dx} \quad (5.8)$$

The MTF is usually defined as the magnitude of this transfer function as a function of positive spatial frequencies. This is a general recipe for finding the MTF of a system element provided that element's LSF is known. The integrals shown in Eq. (5.8) can be reduced to a simple sinc function.

5.3.2 Modulation Transfer Function due to Focal Spot Size

Due to finite focal spot size of the x-ray source in the presence of magnification a blurring is introduced in the images. This is usually referred to as geometrical unsharpness. Geometrical unsharpness increases with the increase in magnification and the effect is severe when the focal spot size increases. Using micro focal x-ray tube this kind of blurring can be minimized to a great extent, which allows for magnification radiography. Fig. 26 shows the geometrical unsharpness due to finite spot size of the x-ray source.

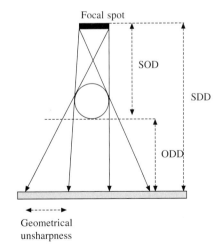

Fig. 26 Geometrical unsharpness due to finite focal spot size. SDD: Source to detector distance, ODD: Object to detector distance and SOD: Source to object distance.

The image of a point in the object space is a spread in the image plane when there is a magnification present in the system and if the source is of finite size. In real life no source is an ideal point source and hence a spread is always associated with the imaging system arising solely due to spot size. This spread is another factor which affects the resolution capability of the imaging system. We can thus associate a MTF due to the focal spot only. Hence, an imaging system can have several MTF associated to different components of the system. The total image system MTF is a product of the MTF of its subsystem. Fig. 27 shows a schematic of the LSF associated with the focal spot.

LSF(x) associated with the intensity distribution due to geometrical unsharpness is the derivative of the ESF(x). In this case the LSF of the focal spot is found by imaging a step (edge) with arbitrary magnification $m = (d_1 + d_2)/d_1$. We can theoretically assume the focal spot to be of rectangular shape and derive an expression for the MTF due to spot size and magnification while assuming the detector is ideal. We can express the LSF due to the spot size as:

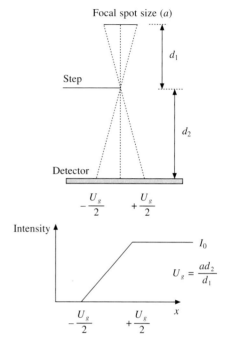

Fig. 27 Schematic of the focal spot blurring process due to a step. d_1 and d_2 are the source to object distance and object to detector distance, respectively.

$$\frac{dI}{dx} = \frac{1}{\left(\frac{U_g}{2} - \left(-\frac{U_g}{2}\right)\right)} = \frac{1}{U_g} \quad (5.9)$$

$$\text{LSF}(x) = \frac{1}{U_g} \quad \text{for } |x| \leq \frac{1}{2} U_g \quad (5.10)$$

$$\text{LSF}(x) = 0 \quad \text{for } |x| > \frac{1}{2} U_g \quad (5.11)$$

Using Eq. (5.8) we can now compute the MTF as

$$M(k) = \frac{\int_{-\infty}^{\infty} \text{LSF}(x) \cdot e^{-ikx} dx}{\int_{-\infty}^{\infty} \text{LSF}(x) dx}$$

$$M(k) = \frac{\int_{-\frac{U_g}{2}}^{\frac{U_g}{2}} \frac{1}{U_g} e^{-ikx} dx}{\int_{-\frac{U_g}{2}}^{\frac{U_g}{2}} \frac{1}{U_g} dx} = \frac{\sin(\pi f_x U_g)}{\pi f_x U_g} = \text{sinc}(f_x U_g)$$

$$M(k) = \text{sinc}(f_x U_g) = \text{sinc}\left(f_x a \frac{d_2}{d_1}\right) = \text{sinc}(f_x a(m-1)) \tag{5.12}$$

A typical plot of Eq. (5.12) can be visualized as a curve as shown in Fig. 28. The curve is meaningful up to the point when it first intersects with x-axis (first zero in y-axis). Using the expression in Eq. (5.12) we can model a MTF response for a focal spot of any size and rectangular distribution at a specific magnification.

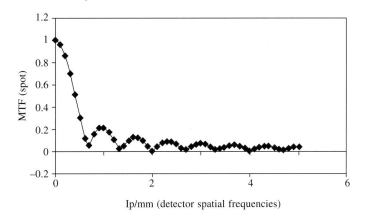

Fig. 28 Typical MTF response due to a finite focal spot size at a certain magnification.

5.4 Detector Quantum Efficiency (DQE)

Although MTF of a system can be measured accurately it cannot be used as the performance metric of an imaging system. The reason is, MTF considers only modulation in different spatial frequencies and is usually measured using high contrast objects under ideal conditions. It has no reference to noise in the system. Noise is inherent in any imaging system and in radiation imaging such as x-ray imaging the radiation source itself is a Poisson random distributed source in the temporal domain. Due to this nature of the source we see a random variation component in the imaging system and this is usually the dominant noise source in modern day digital x-ray imaging system. Detectability of a feature in an image is related to resolution, contrast and noise present in the finally acquired image. While MTF addresses two of these, it has got no measure for noise in the system. Since noise due to quantum fluctuation of the photons coming from the source is usually the dominant noise source, one can get control over this kind of noise by using a very high flux. Increasing the photon flux increases the mean of the response at the detector and being Poisson in nature the associated standard deviation relatively decreases with the increase in mean. In a typical Poisson statistics the mean of the distribution is proportional to the standard deviation of the distribution. Increase in flux reaching the detector can be increased either by increasing the flux in the x-ray tube itself or by increasing the exposure time. In digital x-ray imaging one can also use frame averaging in the temporal domain by acquiring several frames, if long exposure time is a limitation on the system.

Noise in the system is, therefore, related to flux levels and hence to dose irradiated. In a medical imaging scenario one of the major objective is to reduce dose to the patients and obtain better image quality. Similarly, in an industrial scenario the objective is to reduce dose to the detector to extend the life of scintillator material and other semiconductor materials and improve productivity. Hence, the demand from both medical and industrial world is to work with minimum dose and as fast as possible. Under such circumstances the noise performance of the detector is important. Hence, we need an image quality metric that can address both the resolution capability as well as noise performance of the system at lower radiation dosage.

The detector quantum efficiency (DQE) is therefore introduced in medical imaging and later adapted among industrial community as a suitable image quality metric for evaluating system performance [16-21]. Despite its being widespread, this parameter is not generally well understood and not properly measured. The quantum efficiency (QE) is defined as the average fraction of the input quanta, that is used in the formation of the output signal. DQE is generally defined as the ratio of the squared output signal-to-noise ratio (SNR) to the squared input SNR of the imaging detector [16]. Assuming a quasi-ideal imaging detector we will first derive an expression for the DQE in the spatial domain. The quasi-ideal imaging detector only has noise due to the input Poisson statistics and the fluctuations introduced by the binomial statistics since QE < 1 [21]. An x-ray quantum on the scintillator interacts (probability = QE) or does not (probability = (1–QE)). Therefore, this is a binary selection process. According to the binomial theorem the variance of QE can be expressed as QE (1 – QE).

$$\sigma_0^2 = (\sigma_i)_0^2 + \sigma_{QE}^2 \tag{5.13}$$

$(\sigma_i)_0^2$ represents the input noise variance as viewed at the detector output and σ_{QE}^2 is the variance introduced by the binomial fluctuation (due to attenuation in the scintillator).

$$(\sigma_i)_0^2 = (QE \cdot \sigma_i)^2 = (QE)^2 \cdot (\sigma_i)^2 = (QE)^2 \cdot \overline{S}_i \tag{5.14}$$

$$(\sigma_{QE})^2 = QE \cdot (1 - QE) \cdot \overline{S}_i \tag{5.15}$$

$$\sigma_0^2 = (QE)^2 \cdot \overline{S}_i + (QE) \cdot (1 - QE) \cdot \overline{S}_i = (QE) \cdot \overline{S}_i = \overline{S}_0 \tag{5.16}$$

where $(QE) \cdot \overline{S}_i$ is the average output signal and $\overline{S}_0$ represents the variance due to the Poisson statistics. If QE = 1, then $\sigma_0^2 = \sigma_i^2 = \overline{S}_i = \overline{S}_0$.

By introducing square of SNR in the expression we can derive a much simpler definition for the QE.

$$\left(\frac{\overline{S}_0}{\sigma_0}\right)^2 = \frac{(QE)^2 \cdot \overline{S}_i^2}{(QE) \cdot \overline{S}_i} = (QE) \cdot \frac{\overline{S}_i^2}{\sigma_i^2} \tag{5.17}$$

$$QE = \frac{\left(\dfrac{\overline{S}_0}{\sigma_0}\right)^2}{\left(\dfrac{\overline{S}_i}{\sigma_i}\right)^2} = \frac{\overline{S}_0}{\overline{S}_i} \tag{5.18}$$

Now, in the case of a real imaging detector, there is expected to be one more noise source, that is, any adjunctive noise sources (variance σ_n^2). This noise source is also uncorrelated with the Poisson and the Binomial distribution sources:

$$\sigma_0^2 = (\sigma_i)_0^2 + \sigma_{QE}^2 + \sigma_n^2 \tag{5.19}$$

Using Eq. (5.16)

$$\sigma_0^2 = \overline{S}_0 + \sigma_n^2 \tag{5.20}$$

This indicates the real detector will have smaller $(SNR)_0$ than that of a quasi-ideal detector. However, the real detector has the same QE as a quasi-ideal detector. Hence, Eq. (5.18) cannot represent QE of a real detector (in the presence of σ_n^2). It represents the QE of the equivalent quasi-real detector. This equivalent QE is the DQE.

$$\text{DQE} = \frac{\left(\dfrac{\overline{S}_0}{\sigma_0}\right)^2}{\left(\dfrac{\overline{S}_i}{\sigma_i}\right)^2} = \frac{\overline{S}_0^2}{\dfrac{\overline{S}_0 + \sigma_n^2}{\overline{S}_i}} = \frac{(\text{QE})}{1 + \dfrac{\sigma_n^2}{(\text{QE}) \cdot \overline{S}_i}} \tag{5.21}$$

The DQE of the detector, hence mostly related closely to the QE of the detector at high input quanta and at low value of input quanta, the adjunctive noise of the system plays a role.

DQE can be also expressed in the frequency domain using the power spectral density which represents the average power of the signal or noise at a given spatial frequency.

$$\text{DQE} = \frac{\left(\dfrac{\overline{S}_0^2}{\overline{S}_0 + \sigma_n^2}\right)}{\left(\dfrac{\overline{S}_i}{S_i}\right)^2} = \frac{\dfrac{(\text{QE})^2 \cdot \overline{S}_i^2}{(\text{QE}) \cdot \overline{S}_i + \sigma_n^2}}{\left(\dfrac{\overline{S}_i^2}{S_i}\right)} \tag{5.22}$$

Eq. (5.21) can be further simplified as

$$\text{DQE}(f) = \frac{\dfrac{(\text{QE})^2 W_i(f) \text{MTF}(f)^2}{(\text{QE}) \cdot W_{ni}(f) \text{MTF}(f)^2 + W_n(f)}}{\dfrac{W_i(f)}{W_{ni}(f)}} \tag{5.23}$$

where $W_i(f)$ is signal power spectrum, $W_{ni}(f)$ the noise power spectrum at the input of the detector and $W_n(f)$ is the power spectrum of the adjunctive noise. The factor $\text{MTF}(f)^2$ permits the filtering of the power spectrum for the signal and noise through the system transfer function. Eq. (5.23) has been simplified by many authors to different simplified versions [16]

$$\text{DQE}(f) = \frac{\overline{d}^2 \cdot \text{MTF}(f)^2}{\overline{q} \cdot \text{NPS}(f)} \tag{5.24}$$

where $\overline{d}$ is the average pixel value in a "flat-field" dark-subtracted image (calibrated) and it is unitless. MTF(f) is the modulation transfer function (unitless), $\overline{q}$ is the average density of x-ray quanta incident on the system (mm^{-2}) and NPS is the noise power spectrum measured in mm^{-2}.

Since dose is a measure of noise and MTF is a measure of the contrast transmittance, DQE can be visualized in a much simpler form as the ratio between the SNR and dose falling on the detector. From an image standpoint, SNR can be replaced with contrast-to-noise ratio (CNR). Hence, CNR per dose is a relative measure of DQE, i.e.

$$DQE = \frac{CNR}{Dose} \tag{5.25}$$

5.5 Image Quality in the Image Domain

Once the data is acquired, the image is fixed and recorded in the memory. For convenience usually CNR is used to quantify the recorded image quality. We expect as high as possible CNR value to bring out the features in the images clearly and separate the noise level. The stochastic nature of image quanta imposes a fundamental limitation on the performance of photon-based imaging systems. This was first recognized by Rose [22-24] in 1948 and his contemporaries [25-27]. The task is to determine whether or not an object is present in an x-ray image. Rose criterion is a method that allows one to estimate how well a human being can perform the task of identifying an object from the background in an image.

Assume a small uniform circular disk as the object as shown in Fig. 29. When imaged onto a detector using x-rays there will be a higher signal in the background and a lower signal in the circular object. The average fluences are represented as Φ_{BG} and Φ_{OB}, respectively (photons/pixel).

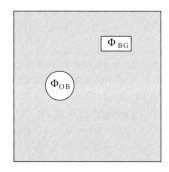

Fig. 29 Model x-ray image of a circular disk.

The contrast C of the object can be defined as the ratio of the difference between the mean x-ray photon fluence in the image of the object Φ_{OB} and that outside the object Φ_{BG}, and Φ_{BG}, i.e.

$$C = (\Phi_{OB} - \Phi_{BG})/\Phi_{BG} \tag{5.26}$$

We can compute the area A of the object as

$$A = \pi R^2 \tag{5.27}$$

We can also find in the radiograph the circular region of radius R and therefore compute from the image, how many photons N actually fell in this circular region.

$$N_{OB} = A * \Phi_{OB}$$

$$N_{BG} = A * \Phi_{BG}$$

where N_{OB} and N_{BG} are the number of photons in the object and background area, respectively. We can now define a threshold value, N_{TH} (approximately) half-way between N_{OB} and N_{BG}

$$N_{TH} \approx (N_{OB} + N_{BG})/2 \tag{5.28}$$

If N is greater than N_{TH} the machine decides that the object is present in the image.

The detectability of the object will depend on the statistical variability associated in N. If N has a Poisson distribution, it will have a standard deviation σ = sqrt (average value of N). For example, in the model since we are considering a x-ray image the photons arrival follows Poisson statistics and

$$\sigma \approx \text{sqrt}(A * \Phi_{BG}) = \text{sqrt}(N_{BG}) \tag{5.29}$$

Since N_{BG} is known we can compute σ. If $\sigma \ll |N_{BG} - N_{OB}|$ we can easily detect the object in this image. If $\sigma \gg |N_{BG} - N_{OB}|$ the detectability of the object will be less. It is natural, therefore, to define a contrast-to-noise ratio (where the contrast is the incremental difference in the number of photons due to the presence of the object) as:

$$K = |N_{BG} - N_{OB}|/\sigma \tag{5.30}$$

Rose's criterion states that the threshold of detectability of the object for human will be for $K = 2$ to 5. The criterion is independent of x-ray images or on the uniformity of the object to be detected. This human threshold value of K for detectability seems independent of the type of system that is used to produce the images. Rose's criterion is a biophysical hypothesis and not a mathematically derived formula.

5.6 Noise in x-ray Images

In a digital x-ray imaging system there are variety of sources for noise. Noise is contributed from the x-ray source, inspected object, image recording device, associated electronics and also the environment. Noise has a serious influence on the imaging system. Various techniques have been developed to suppress noise and improve the CNR for acquired image data. The noise transfer characteristics of imaging systems are strongly influenced by the number of quanta of energy propagating through each step of the imaging system. An insufficient number of quanta at a particular stage lead to a process called as quantum sink. The propagation of image quanta through different stages of the imaging chain (either a gain or spreading process) may be plotted schematically in a spatial frequency dependent quantum accounting diagram [28], sometimes referred to as QAD. The QAD plots the product of the gains and squared MTF at each stage in the system normalized to unity at stage 0. These QAD plots can be quickly referred to qualitatively understand the transfer characteristics of the imaging chain. A typical QAD curve is shown in Fig. 30. This can quickly convey the message as to what stage and at which frequency a quantum sink is occurring.

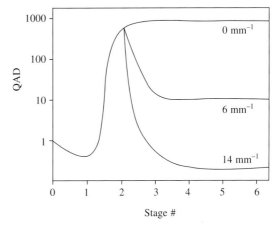

Fig. 30 A typical QAD curve for various stages and at different spatial frequencies.

The stages (or components of an imaging chain) in a typical digital radiography system are as follows [28]. Stage 0: transmitted x-ray spectrum falling on the detector (Poisson distribution) stage 1: absorption of x-rays in the converting scintillator material (Binomial distribution), stage 2: formation (Poisson distribution) and emission (Binomial distribution) of optical photons within the scintillator material, stage 3: spread of these visible photons inside the scintillator material, stage 4: coupling of the optical photons with the photo diodes (Binomial distribution), stage 5: integration of the quanta by the photo diode (deterministic step).

There is also a variety of electronic additive noise in the system. Mainly four sources of additive noise are considered, viz. (1) intrinsic noise from the a-Si pixel, (2) noise due to voltage fluctuations on the gate and bias lines, (3) noise from the amplifier and (4) digitization noise.

Since each of these noise sources is statistically independent of each other, the variance can be added for the total effective noise calculations.

Overall noise characteristics observed from an a-Si-based flat panel detector system shows the noise to be quantum limited and follows a Poisson distribution. This indicates that photon noise constitutes the major source of noise for this imaging system.

6. Selection Criteria of Detectors for Industrial Applications

We have discussed various types of digital x-ray imaging systems used for NDE applications. X-ray image intensifier, computed radiography using storage phosphors, 2D array detectors: a-Si, a-Se, CMOS and CCD and linear diode arrays are few examples of imaging devices. Understanding the physics, advantages and limitations of various systems would help the users to choose the appropriate imaging device for his intended application.

Selection of imaging systems is based on intended application. The requirements for field radiography are portable and flexible detectors, good imaging performance (example: with isotope sources), capability to withstand harsh environments, shorter exposure time and meet image quality requirement as required. Film radiography and computed radiography would be suitable for these types of applications. However, CR has advantages of shorter exposure time, faster image acquisition without chemical processing and direct digital image. Corrosion detection of process pipes in petrochemical industries is an example where computed radiography can be more beneficial.

On-line inspection of castings for high productivity is the requirement in automotive industries. This needs systems with faster acquisitions time, larger dynamic range and higher DQE. XIIs can be used for these applications. However, 2D a-Si array detectors (static or real-time) would be more beneficial in terms of image quality requirements. CR can be more beneficial for inspecting castings off-line in foundries because of its speed and dynamic range over film radiography provided the image quality requirements are met.

Volumetric computed tomography is an application where the 3D image can be acquired in a shorter time. This application needs detectors with faster acquisition time, good spatial resolution and high DQE. 2D a-Si array detectors with real-time capability would be more beneficial for this application. However, XIIs can also be used compromising image quality. Static 2D a-Si array detector can be used for high-resolution images but need to compromise on speed of image acquisition.

CR and DR will play a major role in aerospace industries for inspecting airframes and composite structures because of its speed and direct conversion to digital images. It is essential to understand the key business drivers going for a digital imaging system such as throughput, expected cost savings, standards to be followed and safety requirements. Business drivers can be further flow down to technical requirements such as material, thickness range, geometry, shape, image quality requirements, imaging techniques (2D or 3D), and cycle time needs. We also need to understand the various constraints of the business such as investment plan and operational cost involved. The high level flow chart of system selection is shown in Fig. 31.

The technical requirements can be used for deciding system specifications. The system specifications include DQE, total system MTF, artifacts reduction, dynamic range, operational x-ray energy range, size, weight, operating conditions, flexibility, static/real-time, x-ray source specifications, image review stations requirements and portability requirements.

Surveying the market for various types of digital imaging devices would help the user to make comparisons. An example of qualitative assessment of various imaging devices is given in

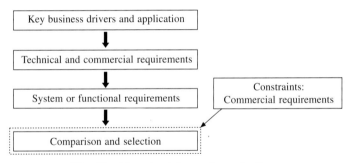

Fig. 31 **Flowchart for system selection.**

Table 3. The best system is that which meets all our requirements and at a lower cost per inspection.

Table 3. Comparison of various types of imaging systems

	Film and Digitization	Image intensifier	CR	Direct conversion	Indirect conversion	CMOS
DQE	***	**	***	****	****	**
Dynamic range	**	**	***	****	****	***
Sensor size/weight	*****	*	*****	***	***	***
Operating conditions	****	***	***	***	***	***
Initial investment	*****	****	***	**	**	**
Througput	*	***	***	****	*****	****
ROI	**	**	***	****	****	***
Flexibility	*****	*	****	*	*	*
Artifacts	****	**	****	****	****	****

7. Conclusion

Various digital imaging options for non-destructive evaluation applications are discussed. From the basics of how various kinds of detectors work, their imaging principle, various associated artifacts and applications are presented in detail. From the basic principle of how a scintillator converts x-ray energy to electrical signal we have attempted to provide a standard for selecting the right kind of imaging options for various applications.

Understanding the image quality quantification is a crucial task before we employ an imaging system for certain inspection. Resolution capability is one of the metric but contrast-to-noise ratio plays an important role for detectability. While resolution has been looked upon as a traditional gage for assessing performance of an imaging system we have considered some of the limitations associated with it specific to x-ray imaging. Radiation dose is important both in medical and industrial applications and noise performance of all detectors more or less depend on the dose irradiated. Better contrast-to-noise ratio, lower dose and faster image acquisition while keeping the cost down are the general requirements for many imaging applications. While it may not be possible to optimize to the ideal conditions of a detector in one system we can always prioritize the requirements and select the best available imaging system for the application in hand.

The text presents a thorough explanation of how things work to where they can be applied. With the various digital recording devices for x-rays available commercially with a wide range of costs and a very high standard of safety regulations it is not an easy task to select the optimum imaging device. This text will help the user to understand digital radiography technology in detail, quantify performance and optimize the inspection process to meet the required standards.

References

1. David, J. Krus, William, P. and Novak, Lou, The SPIE International Symposium on Optical Science, Engineering and Instrumentation, SPIE Vol. 3768, 1999.
2. E. Kotter and M. Langer, Digital radiography for large-area flat-panel detectors, Physics, Springer-Verlag, pp. 2562–2570, 2002.
3. V.V. Nagarkar, T.K. Gupta, S. Miller, Y. Klugarman, M.R. Squillante and G. Entine, Structured CsI(Tl) Scintillators for X-ray Imaging Applications, *IEEE Trans. Nuclear. Science*, **45**, pp. 226–230.
4. High-resolution CMOS imaging detector, High-resolution CMOS imaging detector, Medical Imaging 2001-Physics of Medical Imaging, SPIE Vol. 4320, 2001.
5. J.G. Rocha, N.F. Ramos, S. Lanceros-Mendez, R.F. Wolddenbuttel and J.H. Correia, CMOS X-rays detector array based on scintillating light guides, "Sensors and Actuators A", Amsterdam: Elsevier, pp. 119–123, 2004.
6. J.A. Rowlands, The Physics of Computed Radiography, *Phys. Med. Biol.* **47**, R123–R166 PII: S0031-9155(02)24746-6, 2002.
7. Jihong Wang, X-ray Image Intensifiers for Fluoroscopy, AAPM/RSNA Physics tutorial for residents at the 1999 RSNA scientific assembly.
8. Clifford Bueno, Computed Radiography, Workshop on Digital Radiography, GE Global Research Center, Bangalore, 2005.
9. M.J. Yaffe and J.A. Rowlands, X-ray detectors for digital radiography, *Phys. Med. Biol.*, **42**, pp. 1–39, 1997.
10. Halmshaw, R., Industrial Radiology: Theory and Practice, second edition, Chapman and Hall, UK, 1995.
11. Cunningham, I.A. and Fenster, A., A method for modulation transfer function determination from edge profiles with correction for finite-element differentiation, *Med. Phys.*, **14**, 533–7, 1987.
12. Fujita, H., Tsai, D., Itoh, T., Doi, K., Morishita, J., Ueda, K. and Ohtsuka, A., A simple method for determining the modulation transfer function in digital radiography, *IEEE Trans. Med. Imaging*, **11**, pp. 34–39, 1992.
13. Judy, P.F., The line spread function and the modulation transfer function of a computed tomographic scanner, *Med. Phys.* **3**, pp. 233–6, 1976.
14. Metz, C.E. and Doi, K., Transfer function analysis of radiographic imaging systems, *Phys. Med. Biol.*, **24**, pp. 1079–106, 1979.
15. Reichenbach, S.E., Park, S.K. and Narayanswamy, R., Characterizing digital image acquisition devices, *Opt. Eng.*, **30**, pp. 170–77, 1991.
16. Cunningham, I.A. and Shaw, R., Signal-to-noise optimization of medical imaging systems, *Journal of Optical Society of America-A*, **16**, No. 3/March, pp. 621–632, 1999.
17. Shaw, R., The equivalent quantum efficiency of the photographic process, *Journal of Photographic Science*, **11**, pp. 199–204, 1963.
18. Medical Imaging—The assessment of Image Quality, ICRU report 54, Bethesda, 1995.
19. Shaw, R., Some fundamental properties of xeroradiographic images. In: *Application of Optical*

Instrumentation in Medicine IV, edited by J. E. Gray and W. R. Hendee, Proceedings of SPIE, Vol. 70, pp. 359–363, 1975.
20. Wagner, R.F. and Muntz, E.P., Detective quantum efficiency (DQE) analysis of electrostatic imaging and screen film imaging in mamography. In: *Application of Optical Instrumentation in Medicine VII*, edited by J. E. Gray, Proceedings of SPIE, Vol. 173, pp. 162–165, 1979.
21. Zanella Giovanni, DQE as quantum efficiency of imaging detectors, eprint arXiv:physics, November 2002.
22. Rose, A., Sensitivity performance of the human eye on an absolute scale, *Journal of Optical Society of America*, **38**, pp. 196–208, 1948.
23. Rose, A., Television pickup tubes and the problem of vision. In: *Advances in Electronics and Electron Physics*, edited by Marston, pp. 131-166, Academic Press, New York, 1948.
24. Rose, A., Quantum and noise limitations of the visual process, *Journal of Optical Society of America*, **43**, pp. 715–716, 1953.
25. Fellgett, P.B., On the ultimate sensitivity and practical performance of radiation detectors, *Journal of Optical Society of America*, **39**, p. 970, 1949.
26. Zweig, H.J., Performance criteria for photo-detectors—concepts in evolution, *Photographic Science and Engineering*, **8**, p. 305, 1964.
27. Jones, R.C., A new classification systems for radiation detectors, *Journal of Optical Society of America*, **43**, pp. 715–716, 1953.
28. Siewerdsen, J.H., Antonuk, L.E., El-Mohri, Y., Yorkston, J., Huang, W. and Boudry, J.M., Empirical and theoretical investigation of the noise performance of indirect detection, active matrix flat panel imagers (AMFPIs) for diagnostic radiology, *Medical Physicas*, **24**, No. 1, pp. 71–89, 1997.

Computerized Tomography for Scientists and Engineers
Edited by P. Munshi
Anamaya Publishers, New Delhi, India

16. 3D Tomography Using Neutrons and X-Rays

Amar Sinha and P.S. Sarkar

High Pressure Physics Division, Bhabha Atomic Research Centre,
Trombay, Mumbai-400 085, India

Abstract

Tomography has been widely used in medical diagnosis. In fact, it owes its tremendous development to the widespread demands of medical community. Industrial tomography is a relatively new entry. But unlike medical field, where object to be examined is mostly similar in nature such as human body, industrial tomography poses much greater challenges. An industrial object can be made of a wide variety of complexity and to examine the interiors of such objects, use of different kinds of penetrating radiation are required. For example, neutrons are particularly suited for the examination of hydrogenous materials and to distinguish between isotopes of different materials. X-rays on the other hand are particularly suited for dense objects and distinguishing high and low atomic number objects. This paper presents our work on industrial tomography using neutrons and X-ray. The emphasis is particularly on the technique and methods of 3D neutron and X-ray tomography. In particular, we present two such systems (a) a prototype 3D neutron tomography system developed using APSARA reactor and (b) cone beam 3D X-ray tomography system developed at BARC.

1. Introduction

The last two and a half decades have seen the development of numerous new radiation based imaging techniques for industrial applications [1, 2]. The nature of any industrial object can vary widely and as such requires a wide variety of technique, and probes to be used for their examination. Therefore unlike medical field, several types of penetrating radiation such as X-ray and neutron are being used for probing the interior of an object. Also the tomography methods have been mostly dealing with slice images in what is known as 2D tomography. Though 2D scans do provide useful information of the interior of objects, what is needed is the detailed knowledge of the interior of an object in its true three-dimensional perspective. Such three-dimensional tomography is becoming popular in medical field in the form of spiral CT. In industrial field there is a greater need for such 3D examination as the industrial objects are complex in nature and 2D information alone can be misleading. However, getting such direct 3D information is highly computer intensive and requires special algorithm and instrumentation. That is why in recent years, with the rapid advancement in the field of computers and availability of sophisticated array detectors, 3D tomography has become popular in industrial field also. These 3D tomography methods are not only required to characterize the object under examination but are also being utilized for reverse engineering and rapid prototyping applications. However, 3D tomography is not merely a simple extension of 2D CT scan but requires new algorithm and new imaging techniques associated with sophisticated volume visualization methods to be developed which provides the exact locations and dimensional information of the internal features of the object. These methods are very useful in microtomography, tomography of large objects in a short time and more important for the reverse engineering applications. We describe in this paper a cone beam tomography scanner built in India using a 160 KeV X-ray generator. This paper also describes a prototype neutron tomography system developed using APSARA reactor [3, 4].

2. Neutron Tomography

In this section, development on neutron tomography is discussed along with system description and experimental results. The source distance in the case of reactors is such that the neutron beam can be considered almost parallel when they reach the object under examination. Due to this reason, neutron tomography reconstruction is mostly done in parallel beam geometry. However, the projection data of all the points of the object data, which are illuminated by neutron beam, can be obtained in one single beam exposure. Thus, neutron tomography by its very nature provides projection data of several slices in one single exposure, which makes the 3D reconstruction a straightforward task. We describe the details of the 2D and 3D neutron tomography system.

2.1 Two-dimensional Neutron Tomography

Conventionally, tomography has been carried out using first generation translate-rotate system consisting of photomultiplier based detectors. However, such systems are not especially suited for neutrons as it is very time consuming. It also utilizes a small part of the neutron beam (only a pencil beam) which is otherwise quite costly to produce. We have developed a neutron tomography system [5-9] based on CCD technique and successfully used it at APSARA reactor at Bhabha Atomic Research Centre, for obtaining neutron CT scan of several test specimens and industrial samples. The experimental setup is shown in Fig. 1. Neutron beam coming out of the APSARA reactor after passing through the object under examination falls on the scintillator screen made up of LiF_6 - $ZnS(Ag)$. The flux of neutron falling on scintillator screen is 10^6 n/cm^2/s and the collimation ratio used is 90. This image is acquired through a double stage image intensifier and CCD combination and stored in the PC. The main advantage of CCD based tomography system over conventional translate-rotate tomography is the speed of data acquisition. The object under test is positioned on a sample manipulator and rotated by equal angles through a stepper motor

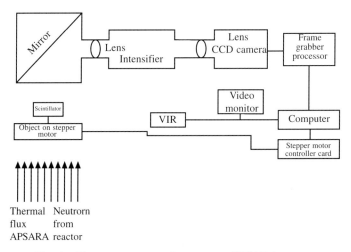

Fig. 1 Neutron tomography setup at APSARA reactor.

drive controlled via software in the PC. These images obtained at each projection angle is preprocessed and subjected to reconstruction algorithm developed in-house for obtaining two-dimensional CT scan images. Fig. 2 shows picture of an object made up of aluminum of 60 mm diameter and containing 9 SS rods and 10 brass rods of 3 mm diameter each. One of the rods is broken halfway. Fig. 3 shows radiograph of the object placed on stepper motor. Fig. 4 (a-c) shows CT scan of this object. We have been able to distinguish between SS and brass rods. This is shown in Fig. 4 (c) by adjusting the contrast and suitably thresholding the CT scan image.

Fig. 2 Aluminum test piece containing 9 SS rods and 10 brass rods (diameter of aluminum piece 60 mm and of rods 3 mm).

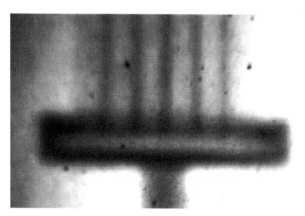

Fig. 3 Radiograph of the object.

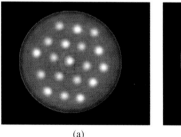 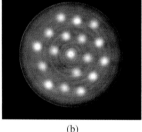 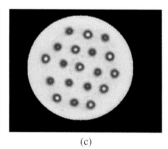

(a)   (b)   (c)

Fig. 4 (a) CT Scan of the object showing 9 SS and 10 brass rods, (b) CT scan of the object at different heights showing one broken rod and (c) Processed CT scan image to highlight differences of attenuation coefficient between SS and brass rods.

2.2 Three-dimensional Neutron Tomography

In recent years, with the advent of CT scan systems capable of giving multiple slices in short time and availability of fast and efficient computers, reconstruction techniques of obtaining 3D tomography images is at the forefront of investigation in medical imaging. It is now possible to interactively examine and see through the interior of an object in three dimensions by combining 2D CT scan data into a 3D volume data. Applications of such techniques for industrial purpose have been rather limited. We have also used this technique of three-dimensional visualization of the interior of industrial objects [10, 11]. There are in general two methods of 3D tomography.

In the first method, three-dimensional tomography is carried out as an extension of the two-dimensional case in which 2D images are stacked one above the other and 3D image is obtained. This form of 3D tomography is usually done where the incoming radiation beam can be considered to be parallel. However, where the source and object distances are such that parallel beam geometry is not valid, a direct 3D cone beam reconstruction is done. This is described in the next section. For the present, we will discuss parallel beam geometry for neutron tomography, as the neutrons coming out of reactor can be considered almost parallel.

In order to demonstrate 3D neutron tomography, two types of phantoms have been used. One of the phantoms made up of several rods of SS and brass inserted inside a 60 mm thick aluminum matrix is already described in Fig. 2. Multiple CT scans of the object shown in Fig. 2 were taken at various heights. Fig. 5(a) shows the object and the horizontal lines show the representative slices at which image reconstruction is done. We have shown only a few such reconstructed CT images in Fig. 5 (b). Fig. 6 shows stacking of these slices using advanced computer technique and the volume rendered images with the opacity of aluminum changed so that it looks almost transparent. Fig. 7 is the cut away view of the same object using cut plane to highlight the broken rod.

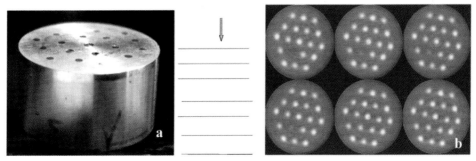

Fig. 5 (a) Aluminum test object containing rods of SS and brass and (b) its multiple CT scan slice images.

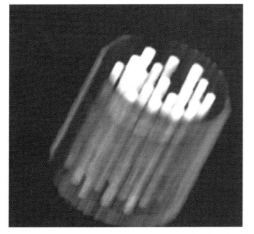

Fig. 6 Volume rendered image obtained using broken stacking of multiple slices shown in Fig. 5(b).

Fig. 7 Cut away view to show broken rod.

The second phantom shown below is a off-centered cone made up of SS embedded inside 40 mm aluminum matrix. This cone has a central hole as shown in Fig. 8 (c). Fig. 8(b) shows radiograph and Fig. 9 the 3D reconstruction of this phantom.

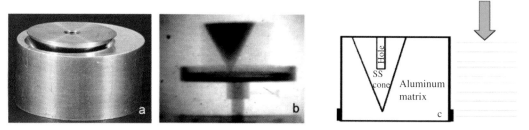

Fig. 8 Example of 3D neutron tomography. Test sample is off-centered SS cone in an aluminum cylinder with a brass ring at the bottom. (a) Picture of the sample, (b) its radiography image along with the sample holder stage and (c) the schematic of the sample and the slices taken at different positions.

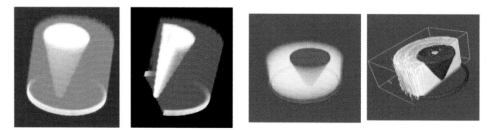

Fig. 9 Reconstructed 3D image of the sample discussed in Fig. 8. Volume rendered images clearly shows the internal off-centered SS cone inside the aluminum cylinder (lightly transparent) and outer brass ring at the bottom end encompassing the Al cylinder.

Thus, the experimental results discussed above demonstrate how the powerful 3D tomography technique helps in examining interior of an object such as position and dimension of the broken rod or the off-centered SS cone in an Al cylinder. The object can be opened layer by layer or cut at any angle using such a visualization technique without physically cutting the material. This is like doing reverse engineering to an object. This has been obtained using prototype tomography setup and only indicative in nature. A full fledged tomography setup is being installed at CIRUS reactor.

3. X-ray Tomography

3.1 Two- and Three-dimensional X-ray Tomography

Fig. 10 shows schematic diagram of a 2D tomography system used in many medical and industrial applications. An X-ray fan from an X-ray source penetrates the object and the attenuation is measured by a linear detector. In medical tomographs, the source and detector rotate around the object (patient). In industrial applications it is in most cases advantageous to rotate the object. During the rotation a set of one-dimensional projections are measured and reconstructed. The result is a two-dimensional image.

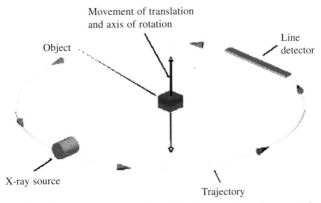

Fig. 10 Schematic of a two-dimensional fan beam X-ray tomography system.

To get a three-dimensional image with such a conventional tomograph, the object has to be moved along the axis of rotation and several scans are performed. A stack of slices are obtained and then using advanced computer algorithm mounted one above the other to get a three-dimensional image. This procedure is very time consuming and corrections also have to be incorporated for the different magnifications in horizontal and vertical directions of the object while reconstructing.

3.2 Three-dimensional X-ray Cone Beam Tomography

To overcome this problem of large time consumed in the data collection in the multiple 2D tomography slices and then obtaining 3D tomograph, cone beam tomography is the new emerging solution. A conical beam from an X-ray source penetrates the object placed on a rotating platform as shown in Fig. 11. The X-ray beam transmitted through the sample is modulated by the attenuation in the sample and forms a two-dimensional projection image on the detector. In order to get all projection images, the object is rotated in predefined steps. The set of projections is then used to reconstruct the 3D tomograph of the object [12, 13].

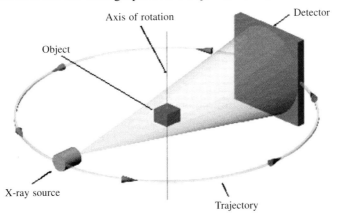

Fig. 11 Schematic diagram of 3D cone beam tomography setup.

In this setup, the detector consists of scintillator, which converts the X-ray signal into visible light. This is then recorded by a cooled digital charge coupled device (CCD) with 16 bit digitization.

The digitized signal which is the 2D projection data is stored for each projection angle and the reconstruction is done using the complex Feldkamp algorithm after all the projections (360°) have been recorded. It provides all projections needed for three-dimensional (3D) reconstruction in a single spin of the object.

Two main families of reconstruction algorithms have been developed for cone beam X-ray tomography with circular source trajectories: the generalized filtered backprojection methods [14] and the 3D Radon transform inversion. Both methods have their own advantages and limitations. We have implemented Feldkamp algorithm for our cone beam tomography setup. This algorithm has been numerically checked using simulated phantoms.

3.2.1 System Description

The overall cone beam tomography experimental setup consists of the following components: 160 kV-4 mA Constant potential X-ray generator with variable focal spot sizes of 1000 μ and 400 μ, a variable mechanical slit assembly to define the size of X-ray beam and shielding to shield the camera from unwanted X-rays and also to limit unwanted scattering, 3 axis (x, y and phi) precise PC controlled sample manipulator with rotation resolution of 0.01°, Gd-Oxysulphide (Tb doped) scintillator, front coated Al glass mirror with reflectivity more than 90%, cooled back-illuminated 1 K × 1 K CCD having pixel size of 13.5 μm × 13.5 μm, cooling to approximately − 50°C, peak quantum efficiency ~90%. The optical coupling is done with 50 mm f/1.2 Nikon make lens and an electronic shutter is employed to control the exposure to CCD. The image is read out and digitized through a MHz controller card (variable readout time 1 to 16 μsec per pixel in 4 steps). After each exposure, depending upon the angular accuracy needed for the specified object reconstruction, the sample is rotated in fixed angular steps till the sample is rotated by a total of 360° from its initial angular position.

3.2.2 The 3D Reconstruction Methodology

The reconstruction is done after the data from all angles have been collected, stored and the region of interest has been identified. The full 3D data indexed by the angle of rotation is at first pre-weighted with the distance between the source to object and object to detector distance and then convoluted in the Fourier space with some filter function. The next step is to backproject each convoluted projection data over a 3D reconstruction grid and then interpolate the data in direct space. We have now with us the reconstructed 3D object details.

3.2.3 Experimental Results

We have performed 3D cone beam tomography experiments [15] with a variety of samples such as reactor fuel bundle end cap, ten turn potentiometer sealed inside a metal casing, fabricated samples like off-centered SS cone in an aluminum matrix etc. These are briefly discussed below.

Fig. 12 shows the photograph of a sample used for 3D reconstruction. This sample consists of end cap of reactor fuel bundle made up of aluminum.

Fig. 13 (a) shows 3D reconstruction of the above sample by taking 200 projections in full 2π. Fig. 13 (b) shows top view of the reconstructed object and Fig.13 (c) shows sliced image of the sample. The experiment was carried out at 120 kV, 4mA, 0.6 sec exposure per scan and − 40°C cooling of CCD.

Fig. 14 (a) shows the radiograph of a sample whose inner and outer material density variation is less than 2% with structures inside. 3D tomography reveals the same whereas the radiograph cannot, as shown in Fig. 14 (b, c).

Fig. 12 Nuclear fuel bundle end cap used for 3D cone beam tomography.

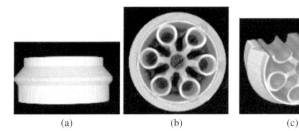

(a)   (b)   (c)

Fig.13 (a) 3D reconstructed volume image of the sample shown in Fig. 12 by taking 200 projections in full 2π, (b) top view and (c) sliced image of the sample.

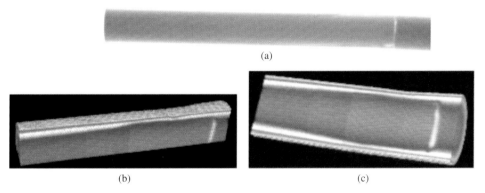

(a)

(b)   (c)

Fig. 14 (a) Radiograph of the sample, (b) and (c) the reconstructed object in 3D (cut-away view): the inner green material and the outer white material has a density difference of < 2%.

3.2.4 System Characterization

We have characterized the system by evaluating parameters such as modulation transfer function of the imaging chain and detection efficiency of individual components of the imaging system.

(a) Modulation Transfer Function (MTF): MTF [16] is the spatial frequency response of an imaging system. The basic idea is to evaluate the resolution performance in optical systems, i.e. its ability to display high resolution images. For this purpose, we have used edge spread function (ESF), which is determined from a line profile perpendicular across the edge. The FFT of the derivative of the line profile is used to estimate the MTF in terms of spatial frequency (lp/mm). Higher spatial frequencies correspond to fine image detail. MTF of the imaging chain was

experimentally calculated using tungsten foil of 200-micron thickness. At 120 kV and 4 mA the radiograph was taken with optimum exposure and the 10% MTF was calculated to be 2.44 lp/mm (= 409 micron). The radiograph (Fig. 15 (a)) of the tungsten foil, its line profile along the horizontal edge (Fig. 15 (b)), the derivative of the profile (Fig. 15 (c)) and the FFT giving the MTF (Fig. 15 (d)) are shown in Fig. 15.

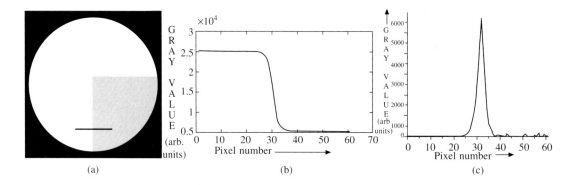

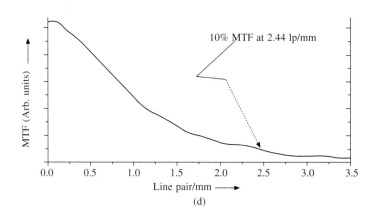

Fig. 15 (a) Radiograph of the tungsten foil, (b) line profile of the edge, (c) derivative of the line profile and (d) estimated MTF of the system with sampling interval of 0.1165 cm.

We have also calculated the MTF of the same imaging chain using a 6 mm thick and 60 mm diameter lead ring and compared with lead lined resolution chart. At 120 kV and 4 mA the radiograph (Fig. 16 (a)) was taken with 1.0 sec exposure and the 10% MTF (Fig. 16 (c)) was calculated to be 2.38 lp/mm.

At the same parameters of the X-ray tube, lead lined resolution chart was imaged. At the point of ~10% difference between the crest and trough gray values, the resolution chart read 2.4 lp/mm. Fig. 17 (a) and (b), respectively, show the radiographic image of the lead lined resolution chart and the line profile.

The exercise of calculating the resolution with lead lined resolution chart has shown that the theoretical and experimental resolutions agree.

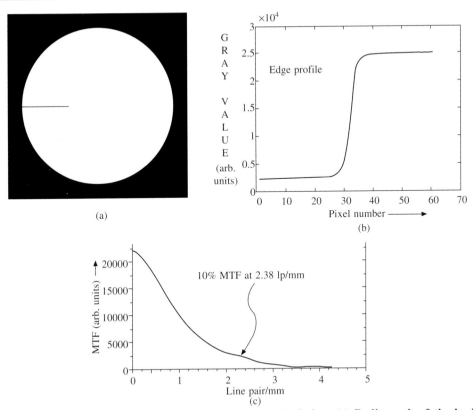

Fig. 16 Calculation of MTF of the imaging chain using lead ring. (a) Radiograph of the lead ring, (b) Edge profile averaged over 4 rows and (C) MTF graph of the imaging chain at 120 kV.

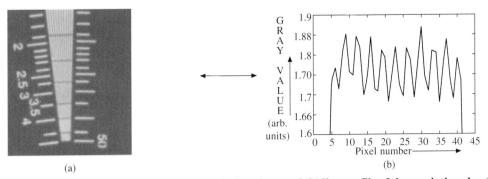

Fig. 17 (a) Radiograph of the lead lined resolution chart and (b) line profile of the resolution chart at 2.4 lp/mm.

We have also calculated the MTF of tomographic system as a whole using an aluminum cylinder of 39 mm diameter. The MTF value was calculated at different positions of the object along the rotation axis from the mid-plane. The variation in the 10% MTF was found out to be around 4.65 lp/mm in the mid-plane to 3.25 lp/mm at an angle of 3.28° (with respect to the mid-plane fan) below the mid-plane. Fig. 18 depicts the MTF (10%) at mid-plane and at a plane away from the mid-plane.

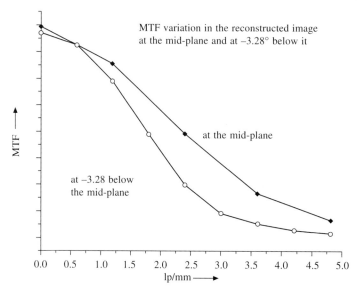

Fig. 18 MTF of the tomographic system at two different positions along the z-direction (axial direction).

As the MTF is a non-isotropic and spatially varying, we have measured the MTF at different positions in the z-direction, i.e. the axial direction. Fig. 19 shows variation of MTF as we move away from the mid-plane, i.e. away from 0°.

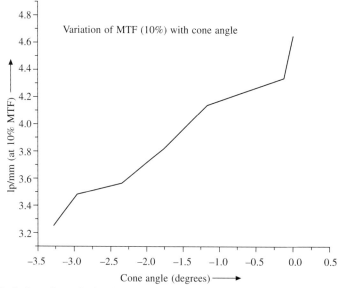

Fig. 19 Variation of resolution in terms of MTF along the z-direction (the axial direction).

(b) Detection Efficiency (DE): Like MTF, detection efficiency (DE) provides physical measurement of the imaging system. It gives the fraction of the incident signal (X-ray for

scintillator or visible light for lens, CCD etc.) transferred efficiently from one component to the other or from input to the output stage of the imaging chain. These measurements are used to determine how well a system captures information within a range of spatial frequencies. Detection efficiencies of some of the components are discussed below.

(i) *Gadolinium oxysulphide (terbium doped) X-ray screen phosphor* : The DE of this screen was experimentally measured using a photo multiplier tube (PMT). The measured DE of the screen at 100 kV tube voltage is ~ 10%.

(ii) *Front coated glass mirror*: Al front-coated glass mirror has the efficiency of reflecting 85% of incident light.

(iii) The scintillator output is in the green range. The DE of the back illuminated CCD chip in this region is ~65% at room temperature and increases to 80% at the working temperature of $-40°C$ ambient. The supplier of the cooled CCD supplied the data.

4. Conclusion and Remarks

A 3D neutron and X-ray tomography system has been developed [17]. The neutron tomography has been developed using APSARA reactor and 3D neutron tomography has been carried out using stacking of 2D slices as the neutron beam coming out of reactor can be considered to be almost parallel. The 3D X-ray tomography has been done using direct cone beam geometry using Feldkamp algorithm. This has been developed using state-of-the-art imaging components. The entire hardware and software for the system is automated. The performance of the system and the reconstruction code has been successfully tested using samples of various contrast and resolution and the results obtained are highly encouraging. Though there are CT systems available using cooled CCD for micro-tomography applications but very few are for general industrial purpose. Our attempt is to develop an industrial 3D tomographic unit with optimum resolution and detail detectability. Though the feasibility of the system has been demonstrated for a X-ray machine of 160 kV, the technique can easily be extended to MeV ranges with suitable choice of hardware. This will make it a very valuable tool for examination of large assembled components such as rockets, missile, fuel bundles etc.

References

1. A. Sinha, Advances in Neutron Radiography Applications, *Journal of Nondestructive Evaluation*, **19**, No. 4, Dec. 1999.
2. Amar Sinha, Digital imaging of neutrons and its application, BARC NewsLetter # 183, April 1999.
3. Amar Sinha and A.M. Shaikh, Experiments with electronic imaging system at APSARA reactor, Presented at the 6th World conference on neutron radiography held at OSAKA, Japan, May 17-21, 1999.
4. Amar Sinha, B.D. Bhawe, G.G. Panchal, A. Shyam and M. Srinivasan, Imaging of nuclear scintillations and its applications, Presented at the Symposium on Nuclear Physics Bhubneshwar, Dec. 26-30, 1994.
5. Amar Sinha, Development of three dimensional neutron tomography system and its applications, *Rev. Sci. Instrum.*, **71**, No. 3, March 2000.
6. Amar Sinha, B.D. Bhawe, C.G. Panchal, A. Shyam, M. Srinivasan and V.M. Joshi, High sensitivity neutron imaging system for neutron radiography with a small neutron source, Nuclear Instrument and Methods, **B 108**, 1996.
7. C.G. Panchal, A. Sinha, B.D. Bhawe, A.Shyam and M. Srinivasan, Digital image processing of images obtained with pu-be (~10^7n/s) neutron source, *Rev. Sci. Instr.*, **67(8)**, August 1996.
8. Amar Sinha, B.D. Bhawe, C.G. Panchal, A. Shyam and M. Srinivasan, Exploratory studies on neutron

radiography with a small neutron source using nuclear scintillation imaging technique, *Nuclear Instrum. Method.*, **A 377**, 1996.
9. Amar Sinha, B.D. Bhawe, C.G. Panchal, A. Shyam and M. Srinivasan, Development of electronic imaging system for neutron radiography with small neutron sources, Paper presented at the Second National Workshop on Neutron Radiography and Gauging, held at Jodhpur, 9 Feb. 1996.
10. Amar Sinha, A.M. Shaikh, and A. Shyam, Development and characterization of a neutron tomography system based on image intensifier/CCD system, *Nuclear Instrum. Method.*, **B 142**, 425, 1998.
11. Amar Sinha, Neutron Tomography, Invited talk presented at the Workshop on utilization of Kamini Reactor held at IGCAR, Kalpakkam, Aug. 30-Sept. 4, 1999.
12. A. Sinha, P.S. Sarkar, Y. Kashyap and B.K. Godwal, Developments in radiation imaging techniques, *Insight*, **45**, 145, Jan. 2003.
13. A.C. Kak and Malcolm Slaney, *Computerized Tomographic Imaging*, IEEE Press, 1988.
14. L.A. Feldkamp, L.C. Davis, and J.W. Kress,. *Journ. of. the Optical Society of America* A, **1** 612, June 1984.
15. P.S. Sarkar, A. Sinha, Y. Kashyap, M.R. More and B.K. Godwal, Development and Characterization of a 3D Cone Beam Tomography System, *Nucl. Instr. Meth.* **A 524**, pp. 377–384, 2004.
16. H.E. Martz et al., *Appl. Radiat. Isot.*, **41**, 10/11, 943, 1990.
17. Amar Sinha, P.S. Sarkar and Y. Kashyap, Applications of digital neutron imaging at BARC (India) using reactor and non reactor sources, *IEEE-TNS*, **52**, No. 1, pp. 305–312, Feb. 2005.

Index

Abels inversion method   46
Adiabatic flow   10
Algebraic reconstruction technique (ART)   89, 118, 122-125, 127-129, 137, 138, 142, 145, 146, 164, 165, 172
Annular flow   11, 14, 17, 42
Atomic Energy Regulatory Board (AERB)   38

Beam hardening
   correction   27, 29, 30, 32, 149, 153-155
   effect   148, 152, 154, 156
Blooming   197
Board of Radiation and Isotope Technology (BRIT)   38
Bone
   imaging   48, 56
   mass   48, 57, 64
   mineral density (BMD)   48, 56, 57, 62-65
Born approximation   55
Bremsstrahlung radiation   111
Broadband ultrasonic backscatter (BUB)   62
Bubble deformation   13
Bubbly flow   11-13, 17, 41

Calcaneal devices   60
Capacitance probe   14, 19, 25
Charge
   coupled device   182, 218
   transfer resistance   23
Chebyshev polynomial   115
Chordal segment inversion (CSI)   42
Churn flow   11, 13, 119
Compound ultrasonic tomography   48
Computational fluid dynamics (CFD)   81, 86, 118
Concentration field   75, 76, 78, 87, 133-138, 140, 142, 143, 145, 146, 158, 159, 169, 170, 172, 174
Cone beam tomography   214, 218-220, 225
Constant phase element (CPE)   23, 26
Contrast-to-noise ratio (CNR)   175, 176, 197, 198, 206-208, 210, 211
Convolution back projection (CBP)   27, 68, 87, 90, 92-95, 98, 133, 137, 142, 145, 146, 158, 164, 165, 170-172, 174

Data
   acquisition system   27, 32, 40, 78, 113, 148, 150
   reduction   95
Detection efficiency (DE)   39, 220, 223, 224
Detector
   quantum efficiency (DQE)   187, 198, 204-207, 209, 210, 212
   array   117, 148-150, 211
Differentially heated fluid layer   87, 89, 105
Diffuse layer capacitance   22
Digital radiography   175-178, 181, 185, 188, 191, 196, 197, 208, 211
Distorted born iterative method   49, 55
Double layer
   capacitance   21-23
   effect   25

Electrical
   capacitance tomography   2, 4, 6-8
   process tomography   75, 79
   resistance tomography (ERT)   75, 76, 78, 81, 82, 84-86
Electron cyclotron emission   106
Eurotherm   135, 167

Fan beam   4, 27, 29, 30, 118-120, 130, 150, 151, 218
Faradaic reaction   21, 23
Faraday rotation   106
Field radiography   191, 209
Film radiography   175, 178, 191, 209
Filtered back projection (FBP)   29, 94, 111, 148, 150
Fourier
   transform   31, 93, 94, 104, 164, 165, 172, 174
   -Bessel expansion   115
Fringe patterns   87, 95, 96, 98, 101, 103
Fringing effect   17, 25
Fusion plasma research   106, 110

Gamma
   densitometry tomography (GDT)   43, 47
   ray attenuation   34, 47, 119, 120, 130, 131
   -ray phase fraction-meter (GRFM)   36, 37, 42-45

Gouy-Chapman-Stern (GCS) model   22

Helmholtz model   22
High energy gamma-ray system   120, 121
Homogeneous mixed flow   17

Image intensifiers   176, 185-188, 197, 211
Impedance technique   10, 12, 25
Impurity content   106, 111
Inner layer capacitance   22
Insulated probe   24, 25
Interferograms   87, 89, 91, 95, 97-99, 101, 103
Interferometry   92, 103, 133, 134, 159
Inverse born approximation (IBA)   48
Inversion method   115, 122
Iterative algorithm   51, 52, 158

KDP crystal   158, 159, 163, 167, 168, 170-173
Kinetic regime   158

Laser shadowgraphic tomography   158
Lead-bismuth-eutectic (LBE)   118, 131
Limiting spatial resolution (LSR)   187, 198, 199
Loss of coolant accidents (LOCA)   34

Mach-Zehnder configuration   92
Mass flow rate   19, 21, 36, 44-46
Mechanical manipulators   148, 149, 155
Medical tomography   1, 4, 111
Microwave interferometer   106, 111
Minimum energy algorithms   131
Modulation transfer function (MTF)   179, 187, 198-200, 202-204, 206, 208, 209, 211, 220-223
Multi-phase flow   1, 34-36, 40-43, 46, 47
Multiplicative algebraic reconstruction technique (MART)   118, 125, 127-131, 147

Needle contact probe   12, 15
Neutron tomography   213-217, 224, 225
Noise power spectrum (NPS)   198, 206
Non-destructive
    evaluation (NDE)   148, 156, 188-190, 209, 210
    testing (NDT)   68, 73, 175
Non-intrusive void-fraction   130
Nuclear magnetic resonance (NMR)   1, 3, 110
Nusselt number   95-98

Optical tomography   3, 111, 146, 147
Optimization   36, 92, 122, 126, 137, 138, 198, 211

Oxidation-reduction process   21

Parallel beam   27, 29, 30, 70, 92, 118, 119, 130, 137, 160, 164, 214, 216
Photo
    multiplier tube (PMT)   39, 120, 190, 224
    stimulated luminescence (PSL)   181, 188
Photodiode   88, 179, 183-185, 208
Plasma
    tomography   110
    -wall interaction   106
Poisson distribution   175, 195, 207-209
Poloidal resolution   112
Position emission tomography   3
Process tomography   1-3, 26, 73, 75, 79
Projection slice theorem   93, 94, 164

Quantitative
    ultrasonic tomography (QUT)   49, 51-54, 56
    ultrasound (QUS)   48, 49, 56-58, 60, 62-66
Quasi-stationary   4, 5
Quick-acting valves (QAV)   35, 36, 40, 44, 45

Radiative interference   150
Radial tomography   87, 98
Radon transform   29, 72, 164, 219
Randles' circuit   23, 24
Rayleigh number   88-90, 96-99, 101, 103, 104, 146
Real time detector   27-30, 179
Reconstruction algorithm   8, 9, 50, 51, 89, 98, 120, 134, 138, 148, 159, 164, 165, 171, 174, 215, 219
Region of interest (ROI)   60, 63, 148, 152, 165, 166, 210, 219
Ring artifact   154
Rushton turbine   76, 78

Scanning geometries   29, 150, 155
Scattered illumination   133
Schlieren tomography   133
Scintillator   39, 178-183, 186, 187, 192, 193, 211, 214, 218, 219, 224
Series expansion   92, 106, 111, 122, 126, 137, 138, 172, 174
Shadowgraph   88, 133, 134, 147, 158, 159, 162, 163, 167-173
Signal conditioning   39, 113
Simultaneous iterative reconstruction technique   123, 124
Slices   4, 5, 7, 8, 68-70, 73, 106, 149, 155, 214-218, 224
Slug flow   9, 11, 13, 17, 40

Smart HART transmitters   36
Spatial resolution   3, 7-9, 81, 111-113, 117, 134, 136, 148, 181, 186-188, 191, 193-196, 198, 209
Spectra physics   91, 136
Static mixer   36, 42, 45, 47
Stray capacitance effect   25
Structured packing   5, 6
Surface tension   10, 88-90

Thermonuclear fusion   107, 108
Three dimensional reconstruction   87, 89, 93, 103, 104, 131, 132, 134, 135, 140, 142, 163, 174
Tokamak fusion   107
Tomogram   27, 54, 55, 78, 80, 81, 112, 150-156
Tomographic
   algorithm   87, 89, 92, 93, 95, 98, 99, 126, 130, 132, 133, 137, 140, 142, 145
   inversion techniques   121
   technique   1-3, 9, 110, 131
Tomography
   algorithm   133, 146, 147
   sensor   6, 8, 80
Transducer   35, 39, 51, 52, 59-61, 63, 66
Transmission tomography   52, 55, 92
Turbine flow meter (TFM)   35, 45, 46
Two phase flow parameter   10, 11, 26

Ultrasonic
   reflection tomography (URT)   48-55
   transmission tomography (UTT)   48-54
Ultrasound bone profile index (UBPI)   62
Uncertainty analysis   95
Uninsulated probe   24

Venturi meter   34-37, 42, 44, 47
Viscosity   10, 34, 75, 82
Void fraction   11, 12, 14-18, 25, 93, 118-121, 130, 131, 164
Voltage-to-current (V/I) converter   39, 40
Volumetric computed tomography (VCT)   176, 177, 188, 192, 209

X-ray
   attenuation coefficient   2
   densitometry techniques   48, 57, 63
   detector   4, 29, 177, 178, 180, 181, 183, 184, 200, 211
   source   4, 6, 27-29, 113, 148-150, 152, 178, 185, 188, 195, 196, 198, 199, 202, 208, 209, 217, 218
   tomography   2-5, 7, 9, 28, 106, 111, 213, 217-219, 224
   tube   4, 71, 175, 195, 196, 204, 221